W0038387

Titles in This Series

211 **Takeo Ohsawa,** Analysis of several complex variables, 2002
210 **Toshitake Kohno,** Conformal field theory and topology, 2002
209 **Yasumasa Nishiura,** Far-from-equilibrium dynamics, 2002
208 **Yukio Matsumoto,** An introduction to Morse theory, 2002
207 **Ken'ichi Ohshika,** Discrete groups, 2002
206 **Yuji Shimizu and Kenji Ueno,** Advances in moduli theory, 2002
205 **Seiki Nishikawa,** Variational problems in geometry, 2001
204 **A. M. Vinogradov,** Cohomological analysis of partial differential equations and Secondary Calculus, 2001
203 **Te Sun Han and Kingo Kobayashi,** Mathematics of information and coding, 2002
202 **V. P. Maslov and G. A. Omel′yanov,** Geometric asymptotics for nonlinear PDE. I, 2001
201 **Shigeyuki Morita,** Geometry of differential forms, 2001
200 **V. V. Prasolov and V. M. Tikhomirov,** Geometry, 2001
199 **Shigeyuki Morita,** Geometry of characteristic classes, 2001
198 **V. A. Smirnov,** Simplicial and operad methods in algebraic topology, 2001
197 **Kenji Ueno,** Algebraic geometry 2: Sheaves and cohomology, 2001
196 **Yu. N. Lin′kov,** Asymptotic statistical methods for stochastic processes, 2001
195 **Minoru Wakimoto,** Infinite-dimensional Lie algebras, 2001
194 **Valery B. Nevzorov,** Records: Mathematical theory, 2001
193 **Toshio Nishino,** Function theory in several complex variables, 2001
192 **Yu. P. Solovyov and E. V. Troitsky,** C^*-algebras and elliptic operators in differential topology, 2001
191 **Shun-ichi Amari and Hiroshi Nagaoka,** Methods of information geometry, 2000
190 **Alexander N. Starkov,** Dynamical systems on homogeneous spaces, 2000
189 **Mitsuru Ikawa,** Hyperbolic partial differential equations and wave phenomena, 2000
188 **V. V. Buldygin and Yu. V. Kozachenko,** Metric characterization of random variables and random processes, 2000
187 **A. V. Fursikov,** Optimal control of distributed systems. Theory and applications, 2000
186 **Kazuya Kato, Nobushige Kurokawa, and Takeshi Saito,** Number theory 1: Fermat's dream, 2000
185 **Kenji Ueno,** Algebraic Geometry 1: From algebraic varieties to schemes, 1999
184 **A. V. Mel′nikov,** Financial markets, 1999

Analysis of Several Complex Variables

IWANAMI SERIES IN MODERN MATHEMATICS

Translations of
MATHEMATICAL
MONOGRAPHS

Volume 211

Analysis of Several Complex Variables

Takeo Ohsawa

Translated by
Shu Gilbert Nakamura

American Mathematical Society
Providence, Rhode Island

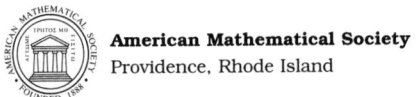

多変数複素解析

TAHENSU FUKUSO KAISEKI
(A MODERN INTRODUCTION TO
SEVERAL COMPLEX VARIABLES)

by Takeo Ohsawa

Originally published in Japanese
by Iwanami Shoten, Publishers, Tokyo, 1998

Translated from the Japanese by Shu Gilbert Nakamura

2000 *Mathematics Subject Classification.* Primary 32Axx.

Abstract. This is an expository account of the basic results in several complex variables that are obtained by L^2 methods.

Library of Congress Cataloging-in-Publication Data

Ohsawa, T. (Takeo)
[Tahensu fukuso kaiseki. English]
Analysis of several complex variables / Takeo Ohsawa ; translated by Shu Gilbert Nakamura.
p. cm. — (Translations of mathematical monographs, ISSN 0065-9282 ; v. 211)
(Iwanami series in modern mathematics)
Includes bibliographical references and index.
ISBN 0-8218-2098-2 (soft cover : acid-free paper)
1. Functions of several complex variables. 2. Mathematical analysis. I. Title. II. Series. III. Series: Iwanami series in modern mathematics.

QA331.7.O3713 2002
515′.94—dc21 2002019351

The paper used in this book is acid-free and falls within the guidelines
established to ensure permanence and durability.
Information on copying and reprinting can be found in the back of this volume.
Visit the AMS home page at URL: `http://www.ams.org/`

10 9 8 7 6 5 4 3 2 1 07 06 05 04 03 02

Contents

Preface

This book does not intend to explain the whole theory of complex analysis in several variables as it stands today. The goal of the book is to introduce methods of real analysis and see these methods produce a variety of global existence theorems in the theory of functions based on the characterization of holomorphic functions as weak solutions of the Cauchy-Riemann equations.

Chapter 1 starts with the definition and elementary properties of holomorphic functions, and in Chapter 2 the problem of extension of functions and the division problem are converted to the problem of solving the Cauchy-Riemann equations of inhomogeneous form. These are called $\overline{\partial}$ equations. The theme to observe up to Chapter 3 is that the solvability of the $\overline{\partial}$ equation on an open set Ω in $\mathbb{C}^n$ imposes on Ω a geometric restriction called pseudoconvexity. Chapter 4 shows, to the contrary, the solvability of the $\overline{\partial}$ equation on a pseudoconvex open set; and, as an application, we generalize to several variables the Mittag-Leffler theorem, Weierstrass' product theorem, and the Runge approximation theorem, which are included in many textbooks for complex analysis in one variable. This approach is called the method of L^2 estimates. By virtue of this method, in Chapter 5, we solve the extension and division problems. The point of this argument is that the solutions are evaluated by the estimates, and thus the application immediately becomes wider. The content stated so far is like the view down from a high place, while Chapter 6 invites the reader to climb the untrodden mountains, which is, so to speak, the view of the great mountains gazed upon from the base camp. The reader will see the author break down from exhaustion there. It is left to the reader to decide whether he has fallen down forward or backward.

Several people offered help with the present publication. In particular, Professor Kazuhiko Aomoto, a great pioneer in analysis, recommended that I write the book and provided useful advice. The editorial board of Iwanami Shoten, Publisher, paid careful attention to the appearance of the book. Mr. Tetsuo Ueda and Mr. Haruo Yokoyama pointed out many mistakes. I am grateful to these people. I regret that just before the approval of the book, my teacher, Professor Shigeo Nakano, who had saved me from dropping out and introduced me to this field, passed away. I would like to offer this book on the altar with my respect. Further, criticism on the book from readers will be considered as my teacher's reprimands from heaven, which I look forward to hearing.

Takeo Ohsawa
May 1998

Preface to the English Edition

Voluminous textbooks have already appeared in several complex variables. In this concise booklet the author assumes a basic knowledge of analysis at the undergraduate level, and gives an account for the L^2 theory of the $\bar{\partial}$ equation. Emphasis is put on recent results which have brought us a deeper understanding of pseudoconvexity and plurisubharmonic functions, and opened a major new way of developing complex analysis.

Summary and Prospects of the Theory

The concept of analytic function was introduced by L. Euler, J. L. Lagrange and others during the 18th century, and it was A. L. Cauchy, C. F. Gauss, G. F. B. Riemann, K. T. W. Weierstrass and others of the 19th century who made the theory of complex functions of one variable as complete and elegant as it is today.

Entering the 20th century, breakthroughs to the world of complex functions of several variables were made by F. Hartogs, E. E. Levi, P. Cousin and others. The problems which they proposed in the field were extremely difficult at the time, but it did not take even a half century to settle all these problems and establish the foundations for the theory of analytic functions of several variables.

As a matter of common knowledge, all the core problems in this area were solved by one mathematician. His name is Kiyoshi Oka (1901-1978).

He grasped all the central problems as a system in the course of solution and gave the last polish to this system by solving affirmatively the so-called Levi problem, which asserts the crucial proposition that if a domain satisfies a geometric condition called pseudoconvexity, one can construct a holomorphic function such that every boundary point of the domain is an essential singularity of this function.

The methods created by Oka were of striking originality. (One of the methods was neatly named "the hovering principle.")

On the other hand, some of the methods contained expressions that were difficult to understand and became obstacles which were hindrances to the succeeding development.

However, it is fortunate that Oka's theorems have been widely accepted today as a lucid fundamental theory, due to H. Cartan's formulation by virtue of cohomology with coefficients in sheaves, and H. Grauert and L. Hörmander's application of methods of functional analysis.

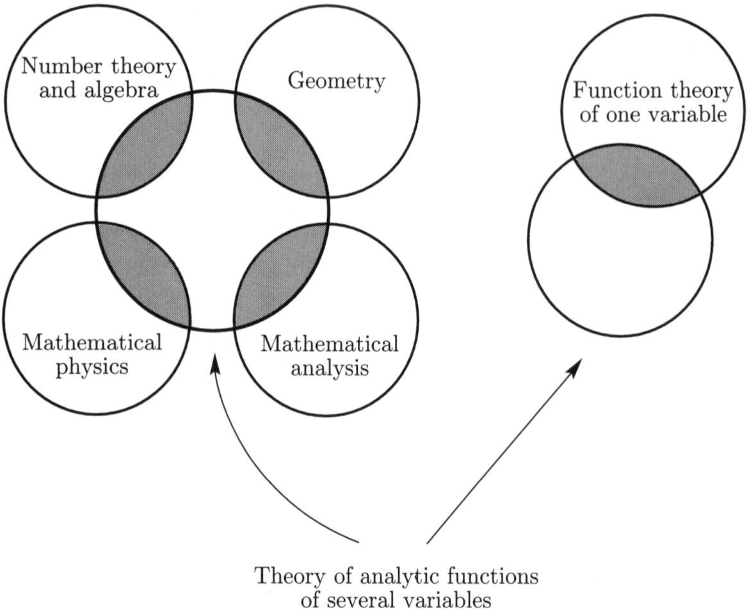

Based on Oka's work, the theory of analytic functions of several variables has developed in a variety of directions. Oka himself suggested how groups of problems should be located by using the schema above.

Borrowing this schema of his, the subjects raised in this book mostly belong to the intersection of the theory of analytic functions of several variables with mathematical analysis. The book hardly touches the other four parts, as noted in the Preface.

As for mathematical analysis, what comes to mind first as an object is those functions dependent on space and time variables that satisfy some sort of differential equations. In order to describe properties of these functions and determine their exact formulae, we need to analyze the distribution of values taken on by these functions and their behavior at points at infinity or singular points, as is often experienced in solving even elementary problems. This methodology forms a complete theory by restricting the range of functions under consideration to analytic ones. This is exactly the theory that was founded by Weierstrass and others within the theory of complex functions of one variable, and the basis for this methodology consists of the Mittag–Leffler theorem, the Weierstrass product theorem, the Runge approximation theorem, and so forth. The Oka theory has had this basis transplanted into the soil of the theory of analytic functions of several variables. On this account, the purpose of this book is to tell the detailed story about how the basis of one variable took root in the soil of several variables, while deferring the sight of what trees grew and blossomed on it.

In what follows, the precise contents of the book are explained.

Chapter 1 gives the definition and fundamental properties of holomorphic functions. P. Montel's theorem and Weierstrass' double series theorem guarantee that the space of holomorphic functions is closed with respect to the topology induced by uniform convergence on compact sets. This is similar to the completeness of the real numbers, and is quite fundamental in deriving existence theorems for holomorphic functions.

From thorough investigation on the structure of the space of holomorphic functions by methods of real analysis, one can derive the fundamental existence theorem. This is grounded on the characterization of holomorphic functions as weak solutions of the Cauchy–Riemann equations in the sense of distributions. Namely, this allows the space of holomorphic functions to be identified with a closed subspace in the space of locally square integrable functions, where functional analytic methods are applicable.

As to individual problems, I. M. Gel'fand, A. Grothendieck and others pointed out that geometric problems of specific spaces can be interpreted as algebraic problems of the function rings on these spaces, and the latter formulation offers an unbroken vista to approach the problems. This relies on the general tendency that the ring structure can be studied well through its ring extensions. From this point of view, the fundamental problem of the ring of holomorphic functions concerns the relation between the set of all the maximal closed ideals of the ring and the domain of definition for the functions. More precisely, the question remains whether 1 belongs to a closed ideal generated by a system of holomorphic functions which have no common zero point, and also whether one can construct a holomorphic function which takes values determined beforehand over a given discrete set. In Chapter 2, these problems are converted into the Cauchy–Riemann equations of inhomogeneous form, or the so-called $\overline{\partial}$ equation. It is too early to treat the $\overline{\partial}$ equation to the fullest, but we introduce the concept of $\overline{\partial}$ cohomology group and make an elementary observation on conditions to solve the above problems. This observation results in a necessary condition that the function ring must contain a function which cannot holomorphically extend beyond a given boundary point of the domain of definition. An open set that has this property is called a domain of holomorphy, and Chapter 3 connects this to a certain geometric concept called pseudoconvexity. The definition and fundamental properties of plurisubharmonic functions, a generalization of subharmonic functions to several variables, are described. This is done by Oka in a 1942 paper based on Hartogs' results. In addition, we prove basic results on the regularization of plurisubharmonic functions, as differentiable plurisubharmonic functions become important later. The approximation theorem due to J.-P. Demailly belongs to the same family of results and hence is introduced here. However, this is rather a deep result, and the proof of the theorem is not elementary and thus is postponed until Chapter 5. We also introduce the concept of Hartogs function, due to Bochner and Martin. This enables us to feel as if we were observing the development of the Levi problem around 1948.

In case an open set in $\mathbb{C}^n$ has a smooth boundary, the pseudoconvexity implies some property of the boundary as a real hypersurface. This is what is called the condition of Levi pseudoconvexity. Open sets of this sort enable one to argue minutely about the boundary behavior of holomorphic functions and mappings. This is a subject of Chapter 6, but Chapter 3 also prepares for this subject.

Chapter 4 explains a new methodology for L^2 estimates of the $\overline{\partial}$ operator. We begin with basics of closed operators on a Hilbert space, establish estimates involving the $\overline{\partial}$ operator and its adjoint operator on the completion of the space of differential forms with respect to the L^2 norm with some weight function, and from these estimates derive the existence theorem on solutions of the $\overline{\partial}$ equation in Theorem 4.11. Further, we apply the theorem and generalize the Mittag-Leffler theorem, the Weierstrass product theorem, and the Runge approximation theorem to several variables. The thread of our argument itself is not different at all from that of Hörmander's book [**27**], the standard textbook in the theory of analytic functions of several variables. But it is worth emphasizing that what is different between our approach and his is the vehicle in which we are traveling, in spite of the same path.

That is to say, the calculations that provide the foundation of the arguments in this book are a re-formation of S. Nakano's formulae used for the proof of the vanishing theorem of cohomology in the theory of complex manifolds, and these calculations contain new estimates. Differing from Hörmander's, in the background of our estimates, there is W. V. D. Hodge's book [**24**] that discusses the theory of harmonic integrals on projective algebraic manifolds, grounded on the symmetry of complex Laplacians with respect to the complex conjugate. It would be of no use if the new vehicle were cheap, but in Chapter 5 by virtue of this new method we show in Theorem 5.10 that holomorphic functions defined on a closed subspace can be extended under an estimate given by the L^2 norm with some weight function. This estimate is what Hörmander's method could not reach, and is the main theme of this book. The calculations of Chapter 4 were, in fact, designed by the author together with Kensho Takegoshi in order to prove this extension theorem. (Refer to [**37**].) The proof of Demailly's approximation theorem is an application of the extension theorem.

Although we cannot treat this topic in this book, the extension theorem has recently shown to have some applicability to some subtle problems in algebraic geometry and the theory of complex manifolds. (Refer to [**44**], [**46**], and [**47**].)

Chapter 5 touches on H. Skoda's division theorem. With this, all the problems posed in Chapter 2 have been solved. We do not give full details of the proof of Skoda's theorem, but only introduce the essential part of the argument. The author would put an emphasis also on this point as a special feature of this book which is not found in any other books.

We throw in Chapter 6 as an extra, "just for fun". It contains a lighthearted approach to the difficult open problem of determining the singularity of the Bergman kernel, which is a reproducing kernel of the space of L^2 holomorphic functions. After describing C. Fefferman and Bell-Ligocka's example of applying the Bergman kernel to holomorphic mappings, we show our own recent results about the Bergman kernel on a general Levi pseudoconvex domain. This might seem too scanty and miserable, but may rouse those who want to do some research in this field from this point forward. That is why the author did not shrink from "cutting a sorry figure".

CHAPTER 1

Holomorphic Functions

To begin with, we define holomorphic functions as convergent power series, describe elementary properties of them, and achieve the main goal of characterizing them as weak solutions of Cauchy-Riemann equations. Our proof restricts itself to locally square integrable functions, while the concept of holomorphy of weak solutions is known to be extendible as far as hyperfunctions. This choice is made so as to keep our argument as simple as possible. At the end, we mention the Reinhardt domain, finding it important to study some properties possessed by domains of convergence for power series.

1.1. Definitions and Elementary Properties

Let $\mathbb{C}$ be the complex plane and consider the n-dimensional complex number space $\mathbb{C}^n := \overbrace{\mathbb{C} \times \cdots \times \mathbb{C}}^{n}$. Let $z = (z_1, \cdots, z_n)$ be the coordinate system of $\mathbb{C}^n$. Essentially z is a vector valued function on $\mathbb{C}^n$, but we also denote a point of $\mathbb{C}^n$ by the same symbol z as long as there is no fear of confusion. Write x_{2j-1} and x_{2j} for the real part $\operatorname{Re} z_j$ and the imaginary part $\operatorname{Im} z_j$ of z_j, respectively. Let $\mathbb{R}$ be the real line, and identify $\mathbb{C}^n$ with $\mathbb{R}^{2n}$ by the correspondence $(z_1, \cdots, z_n) \longmapsto (x_1, x_2, \cdots, x_{2n})$.

For $z \in \mathbb{C}^n$, set

$$|z|_{\max} = \max_j |z_j| \quad \text{and} \quad |z| = (|z_1|^2 + \cdots + |z_n|^2)^{\frac{1}{2}} .$$

These are norms on $\mathbb{C}^n$ that are topologically equivalent to each other.

Let $\mathbb{Z}_+$ be the set of all nonnegative integers, and for an element $\alpha = (\alpha_1, \cdots, \alpha_n)$ in $\mathbb{Z}_+^n$, set

$$\alpha! := \prod_{j=1}^{n} \alpha_j! , \qquad \langle \alpha \rangle := \sum_{j=1}^{n} \alpha_j ,$$
$$z^\alpha := {z_1}^{\alpha_1} \cdots {z_n}^{\alpha_n} .$$

Furthermore, we use the following notation:

$$\begin{aligned}\frac{\partial}{\partial z_j} &:= \frac{1}{2}\left(\frac{\partial}{\partial x_{2j-1}} - \sqrt{-1}\frac{\partial}{\partial x_{2j}}\right),\\ \frac{\partial}{\partial \overline{z}_j} &:= \frac{1}{2}\left(\frac{\partial}{\partial x_{2j-1}} + \sqrt{-1}\frac{\partial}{\partial x_{2j}}\right),\\ \left(\frac{\partial}{\partial z}\right)^{\alpha} &:= \left(\frac{\partial}{\partial z_1}\right)^{\alpha_1}\cdots\left(\frac{\partial}{\partial z_n}\right)^{\alpha_n}.\end{aligned}$$

Also, for $\beta = (\beta_1, \cdots, \beta_{2n}) \in \mathbb{Z}_+^{2n}$, let

$$\left(\frac{\partial}{\partial x}\right)^{\beta} := \left(\frac{\partial}{\partial x_1}\right)^{\beta_1}\cdots\left(\frac{\partial}{\partial x_{2n}}\right)^{\beta_{2n}}.$$

From now on, let Ω denote a nonempty open set in $\mathbb{C}^n$.

DEFINITION 1.1. A complex valued function f on Ω is *holomorphic* if for each point a of Ω, there exists a power series

$$P_a(z) := \sum_{\alpha\in\mathbb{Z}_+^n} c_\alpha (z-a)^\alpha \text{ with } c_\alpha \in \mathbb{C}$$

that converges to f on some neighborhood of a, where the convergence is regarded with respect to some linear order of $\mathbb{Z}_+^n$, or a bijection $\mathbb{Z}_+ \ni k \mapsto \alpha(k) \in \mathbb{Z}_+^n$. f is called *antiholomorphic* if the complex conjugate of f, $z \mapsto \overline{f(z)}$, is holomorphic.

Let $A(\Omega)$ denote the set of all holomorphic functions on Ω.

First, let us describe briefly the most elementary properties of holomorphic functions:

1. $A(\Omega)$ is a $\mathbb{C}$ algebra with the usual four rules of arithmetic. (This is obvious from the definition.)
2. Holomorphic functions are of class C^∞.

In fact, (2) is included in the following proposition:

PROPOSITION 1.2. *If a power series*

$$\sum_{\alpha\in\mathbb{Z}_+^n} b_\alpha (z-a)^\alpha \textit{ with } b_\alpha \in \mathbb{C}$$

around a point $a = (a_1, \cdots, a_n)$ *converges at a point* $a' = (a_1', \cdots, a_n')$ *with* $a_j' \neq a_j$ $(1 \leqq \forall j \leqq n)$, *then for any* $\beta \in \mathbb{Z}_+^{2n}$, *a power series*

$$\sum_{\alpha\in\mathbb{Z}_+^n} b_\alpha \left(\frac{\partial}{\partial x}\right)^{\beta} (z-a)^\alpha$$

converges absolutely and uniformly on compact subsets of

$$\{z \in \mathbb{C}^n \mid |z_j - a_j| < |a'_j - a_j| \text{ for } 1 \leqq j \leqq n\}.$$

In particular, for $f \in A(\Omega)$ *and* $a \in \Omega$,

$$f(z) = \sum_{\alpha \in \mathbb{Z}_+^n} \frac{f^{(\alpha)}(a)}{\alpha!}(z-a)^\alpha$$

on some neighborhood of a, *where* $f^{(\alpha)} := \left(\frac{\partial}{\partial z}\right)^\alpha f$.

PROOF. This follows from the comparison with geometric series. The reader should be familiar with the technique from the theory of complex analysis of a single variable. □

COROLLARY 1.3 (Theorem of Identity). *If* $f \in A(\Omega) \setminus \{0\}$ *and* Ω *is connected, then* $f^{-1}(0)$ *has no interior point.*

PROOF. This follows from the fact that the set

$$\{a \in \Omega \mid f^{(\alpha)}(a) = 0 \text{ for } \forall \alpha \in \mathbb{Z}_+^n\}$$

is both open and closed in Ω. □

When $f \in A(\Omega) \setminus \{0\}$, $a \in \Omega$ and $f(a) = 0$, we call

$$\inf\{\langle\alpha\rangle \mid f^{(\alpha)}(a) \neq 0\}$$

the *order of the zero* of f at a.

Before proceeding further, we explain holomorphic mappings, polydiscs, and complex open balls.

A mapping $F = (f_1, \cdots, f_m)$ from Ω to an open set Ω' in $\mathbb{C}^m$ is said to be *holomorphic* if every component f_j of F is a holomorphic function on Ω. From Definition 1.1 and Proposition 1.2 (in particular, absolute convergence of power series), it follows that the composite of holomorphic mappings is holomorphic. In particular, $\frac{1}{f} \in A\left(\Omega \setminus f^{-1}(0)\right)$ if $f \in A(\Omega) \setminus \{0\}$. A holomorphic mapping $F : \Omega \to \Omega'$ is said to be *biholomorphic* if it has the holomorphic inverse mapping $F^{-1} : \Omega' \to \Omega$. Ω and Ω' are, by definition, *holomorphically equivalent* to each other if there is a biholomorphic mapping from Ω to Ω'. A bijective holomorphic mapping between domains is biholomorphic (see § 5. 4 (a)). Biholomorphic mappings from Ω to itself are called *holomorphic automorphisms.* They form a group with the composition of mappings being the group product. This is called the *holomorphic automorphism group* of Ω and denoted by $\operatorname{Aut} \Omega$.

An *n-dimensional polydisc* around $a = (a_1, \dots, a_n) \in \mathbb{C}^n$ (or simply an *n-polydisc*) is, by definition, a nonempty set

$$\{z \in \mathbb{C}^n \mid |z_j - a_j| < r_j \text{ for } 1 \leqq j \leqq n\},$$

and is denoted by $\Delta(a, r)$, where $r := (r_1, \cdots, r_n)$. For the sake of brevity, we write Δ for $\Delta(0, 1)$. It is clear that $\Delta(a, r)$ is holomorphically equivalent to $\Delta^n := \overbrace{\Delta \times \cdots \times \Delta}^{n}$.

A subgroup of $\operatorname{Aut} \Delta^n$ whose elements fix the origin is isomorphic to a semidirect product of $U(1)^n$ and the n-dimensional symmetric group $\mathfrak{S}_n$. In fact, if $F \in \operatorname{Aut} \Delta^n$ with $F(0, \cdots, 0) = (0, \cdots, 0)$, then the application of (1.2) below to the N times composite $\overbrace{F \circ \cdots \circ F}^{N}$ of F shows that each component of F must be a linear form with respect to z.

Next, for $R > 0$ we call an open set $\{z \in \mathbb{C}^n \mid |z - a| < R\}$ a *complex n-dimensional open ball* around a of radius R (or simply an *n-open ball*) and denote it by $\mathbb{B}(a, R)$. n-open balls are holomorphically equivalent to each other. We express $\mathbb{B}((0, \cdots, 0), 1)$ by $\mathbb{B}^n$ for short.

As in the case of $\operatorname{Aut} \Delta^n$, by knowing that elements of $\operatorname{Aut} \mathbb{B}^n$ that fix the origin are linear, we obtain

$$\{\sigma \in \operatorname{Aut} \mathbb{B}^n \mid \sigma(0) = 0\} = U(n).$$

If $n \geqq 2$, clearly $U(n)$ is not equal to the semidirect product of $U(1)^n$ and $\mathfrak{S}_n$. From this, it is seen that in general Δ^n and $\mathbb{B}^n$ are not holomorphically equivalent to each other[1].

$\operatorname{Aut} \mathbb{B}^n$ acts on $\mathbb{B}^n$ transitively. To see this, it is sufficient to note that $\mathbb{B}^n$ is holomorphically equivalent to the open set

$$D_n := \left\{ \zeta = (\zeta_1, \cdots, \zeta_n) \in \mathbb{C}^n \;\middle|\; \operatorname{Im} \zeta_1 > \sum_{j=2}^{n} |\zeta_j|^2 \right\}$$

under the *Cayley transformation*:

$$z_1 = \frac{\zeta_1 - \sqrt{-1}}{\zeta_1 + \sqrt{-1}}, \; z_2 = \frac{2\zeta_2}{\zeta_1 + \sqrt{-1}}, \cdots, z_n = \frac{2\zeta_n}{\zeta_1 + \sqrt{-1}}.$$

---

[1] In general, the group of holomorphic automorphisms on a bounded domain is known to be a Lie group with respect to the compact open topology (see [**33**]).

In fact, the transitivity of $\operatorname{Aut}\mathbb{B}^n$ is obvious from the facts that $U(n) \subset \operatorname{Aut}\mathbb{B}^n$ and $\operatorname{Aut} D_n$ contains the transformations:

$$\zeta \longmapsto (t\zeta_1, \sqrt{t}\,\zeta_2, \cdots, \sqrt{t}\,\zeta_n) \text{ for } t > 0,$$
$$\zeta \longmapsto (\zeta_1 + t, \zeta_2, \cdots, \zeta_n) \text{ for } t \in \mathbb{R}\,.$$

One of the holomorphic automorphisms on $\mathbb{B}^n$ that map a point $w \neq (0, \cdots, 0)$ to the origin is given by the formula

$$z \longmapsto \frac{\sqrt{1-|w|^2}\left(z - \dfrac{\langle z, w\rangle}{|w|^2} w\right) - w + \dfrac{\langle z, w\rangle}{|w|^2} w}{1 - \langle z, w\rangle},$$

where $\langle z, w\rangle := \sum\limits_{j=1}^{n} z_j \overline{w}_j$.

It is an interesting task to calculate the groups of holomorphic automorphisms on various bounded domains, but we do not study more than these examples in this book.

Let us return to the exposition of elementary properties of holomorphic functions. We wish to describe topological properties of the ring $A(\Omega)$ of holomorphic functions. For this purpose, it is necessary to clarify the range in which the Taylor expansion of a holomorphic function on Ω is convergent.

An immediate consequence from Proposition 1.2 is that the sum of a power series that converges on $\Delta(a,r)$ is a holomorphic function on $\Delta(a,r)$. The converse statement to this does not hold for real analytic functions in general, and this difference between holomorphic and real analytic functions is fundamental.

PROPOSITION 1.4. *A holomorphic function on $\Delta(a,r)$ is equal to the sum of a power series whose convergence holds on the same domain $\Delta(a,r)$.*

PROOF. It is enough to show the proposition for $\Delta(a,r) = \Delta^n$. Let

$$f(z) = \sum_{\alpha} c_\alpha z^\alpha \text{ with } c_\alpha = \frac{f^{(\alpha)}(0)}{\alpha!}$$

be the power series expansion of an element f in $A(\Delta^n)$ at the origin. Since there exists a positive number ε such that the power series converges for $|z|_{\max} < \varepsilon$, c_α is expressed by

$$(1.1) \quad c_\alpha = \frac{1}{(2\pi)^n} \int_0^{2\pi} \cdots \int_0^{2\pi} f(re^{i\theta_1}, \cdots, re^{i\theta_n})\, r^{-\langle\alpha\rangle} e^{-i\langle\alpha,\theta\rangle}\, d\theta,$$

where r is an arbitrary real number with $0 < r < \varepsilon$, and we set $\theta := (\theta_1, \cdots, \theta_n)$ and $d\theta := d\theta_1 \cdots d\theta_n$.

However, since f is holomorphic on Δ^n, it follows that the right hand side of (1.1) takes on finite determinate values for all r with $0 < r < 1$, and is of class C^ω with respect to r. Hence, (1.1) holds for all r with $0 < r < 1$.

From this we obtain the estimate of c_α:

$$|c_\alpha| \leqq \sup_{|z|_{\max}=r} |f(z)|\, r^{-\langle\alpha\rangle} \text{ for } 0 < r < 1. \tag{1.2}$$

Therefore, $\sum_\alpha c_\alpha z^\alpha$ converges on Δ^n. □

The estimate (1.2) implies a fundamental fact of the topology of $A(\Omega)$:

THEOREM 1.5 (Montel's Theorem). *A sequence of holomorphic functions on Ω that is uniformly bounded on compact subsets of Ω contains a subsequence that converges uniformly on compact subsets of Ω.*

PROOF. Let $\{f_k\}_{k=1}^\infty \subset A(\Omega)$ be a sequence in the theorem. For an arbitrary $\Delta(a, r) \Subset \Omega$ (relatively compact), the uniform boundedness on $\overline{\Delta(a, r)}$ implies that

$$M := \sup\{|f_k(z)| \mid z \in \Delta(a, r) \text{ for } k = 1, 2, \cdots\} < \infty.$$

From (1.2), it follows that for any $j = 1, \cdots, n$,

$$\sup_k \left|\frac{\partial f_k(z)}{\partial z_j}\right| \leqq \frac{2M}{r} \text{ for } z \in \Delta\left(a, \frac{r}{2}\right).$$

Therefore, the Ascoli-Arzelà theorem concludes that $\{f_k\}_{k=1}^\infty$ has a subsequence converging uniformly on compact subsets of Ω. □

THEOREM 1.6 (Weierstrass' Double Series Theorem). *If a sequence of holomorphic functions on Ω converges uniformly on compact sets in Ω, then the limit function is holomorphic on Ω as well.*

PROOF. If $f_k \to f$ uniformly on compact sets in Ω and $f_k \in A(\Omega)$, then, as in the proof of Montel's theorem, for any $\alpha \in \mathbb{Z}_+^n$, there exists $f^{(\alpha)}$ such that $f_k^{(\alpha)} \to f^{(\alpha)}$ uniformly on compact sets in Ω. Also, if we take $\Delta(a, r) \subset \Omega$, then the application of (1.2) to $\Delta(a, r/2)$ and the uniform convergence of f_k to f on $\Delta(a, r/2)$ show that for any $\varepsilon > 0$, there exists a positive integer N such that, for any $z \in \Delta(a, r/2)$,

$$\left| \sum_{\langle\alpha\rangle \geqq N} \frac{f_k^{(\alpha)}(a)}{\alpha!}(z-a)^\alpha \right| < \frac{\varepsilon}{2} \quad \text{for} \quad k = 1, 2, \cdots$$

and

$$|f_k(z) - f(z)| < \frac{\varepsilon}{2} \quad \text{for} \quad k > N.$$

Hence,

$$\begin{aligned} &\left| f(z) - \sum_{\langle\alpha\rangle < N} \frac{f^{(\alpha)}(a)}{\alpha!}(z-a)^\alpha \right| \\ &\quad \leqq |f(z) - f_k(z)| + \left| f_k(z) - \sum_{\langle\alpha\rangle < N} \frac{f_k^{(\alpha)}(a)}{\alpha!}(z-a)^\alpha \right| \\ &\qquad + \left| \sum_{\langle\alpha\rangle < N} \frac{f_k^{(\alpha)}(a) - f^{(\alpha)}(a)}{\alpha!}(z-a)^\alpha \right| \\ &\quad < \frac{\varepsilon}{2} + \frac{\varepsilon}{2} + \left| \sum_{\langle\alpha\rangle < N} \frac{f_k^{(\alpha)}(a) - f^{(\alpha)}(a)}{\alpha!}(z-a)^\alpha \right|. \end{aligned}$$

Accordingly, as $k \to \infty$, it follows that

$$\left| f(z) - \sum_{\langle\alpha\rangle < N} \frac{f^{(\alpha)}(a)}{\alpha!}(z-a)^\alpha \right| \leqq \varepsilon.$$

Therefore, the power series $\sum \frac{f^{(\alpha)}(a)}{\alpha!}(z-a)^\alpha$ converges on $\Delta(a, r/2)$, and its sum is equal to $f(z)$. □

From now on, the origin $(0, \cdots, 0)$ is denoted simply by 0 if there is no fear of confusion.

THEOREM 1.7. *If Ω is connected, nonconstant holomorphic functions on Ω are open mappings.*

PROOF. For $f \in A(\Omega) \setminus \mathbb{C}$ and $a \in \mathbb{C}$, take a holomorphic mapping $h : \Delta \to \Omega$ satisfying $h(0) = a$ and $f \circ h \in A(\Delta) \setminus \mathbb{C}$. Then on some neighborhood U of 0, there exists a decomposition

$$f(h(\zeta)) = \zeta^m g(\zeta) + f(a) \quad \text{for} \quad \zeta \in U,$$

where $m \in \mathbb{N}$, the set of positive integers, and $g \in A(U)$ with $g(0) \neq 0$. Since the mapping $\zeta \mapsto \zeta^m$ is locally a differentiable homeomorphism except for $\zeta = 0$, there is a holomorphic function u satisfying $f(h(\zeta)) - f(a) = (\zeta u(\zeta))^m$ on some neighborhood of 0. Since

$u(0) \neq 0$, the inverse mapping theorem implies that $\zeta u(\zeta)$ has a differentiable inverse on some neighborhood of 0. Therefore, $\zeta u(\zeta)$ is an open mapping, and so is $f(h(\zeta))\,(= (\zeta u(\zeta))^m + f(a))$. Hence, f maps neighborhoods of a to neighborhoods of $f(a)$. □

COROLLARY 1.8 (Maximum Principle). *If Ω is connected, the absolute value of a nonconstant holomorphic function on Ω has no maximum in Ω.*

PROOF. The reasoning is that any open disc of $\mathbb{C}$ contains a point whose distance to 0 is larger than the absolute value of the center of the open disc. □

From this, an important proposition in the theory of holomorphic mappings follows:

THEOREM 1.9 (Schwarz's lemma). *If $f \in A(\Delta^n)$, $f(0) = 0$, and $\sup_{z\in\Delta^n} |f(z)| = M$, then*

$$|f(z)| \leqq M|z|_{\max} \tag{1.3}$$

on Δ^n.

PROOF. For $a = (a_1, \cdots, a_n) \in \partial\Delta^n$, consider a holomorphic mapping

$$\begin{array}{cccc} \pi_a : & \Delta & \longrightarrow & \Delta^n \\ & \Cup & & \Cup \\ & \zeta & \longmapsto & (a_1\zeta, \cdots, a_n\zeta), \end{array}$$

and apply the maximum principle to $f \circ \pi_a(\zeta)/\zeta$. Then we obtain

$$|f \circ \pi_a(\zeta)| \leqq M|\zeta|,$$

where the equality holds if and only if $f \circ \pi_a(\zeta)/\zeta \in \mathbb{C}$. Hence, (1.3) follows from $|\zeta| = |\pi_a(\zeta)|_{\max}$. □

1.2. Cauchy-Riemann Equations

Differential equations

$$\frac{\partial f}{\partial \overline{z}_j} = 0 \text{ for } j = 1, \cdots, n$$

on Ω are called the *Cauchy–Riemann equations.* Holomorphic functions are all solutions of the Cauchy-Riemann equations since $\partial z^\alpha/\partial \overline{z}_j = 0$ for any $\alpha \in \mathbb{Z}_+^n$, and convergent power series are differentiated term by term. The aim of this section is to characterize holomorphic functions as locally square integrable functions that satisfy the Cauchy-Riemann equations.

To begin with, we introduce some notation and terminology.

Let $L^2(\Omega)$ denote the Hilbert space of all complex valued measurable functions on Ω that are square integrable with respect to the Lebesgue measure, where two functions whose values coincide everywhere except on measure zero sets are considered to be the same.

We identify the Lebesgue measure on $\mathbb{C}^n$ with the volume element with respect to the *Euclidean metric*

$$dx_1 \wedge dx_2 \wedge \cdots \wedge dx_{2n} \left(= \left(\frac{\sqrt{-1}}{2} \right)^n dz_1 \wedge d\overline{z}_1 \wedge \cdots \wedge dz_n \wedge d\overline{z}_n \right),$$

and denote it by dV, dV_n, or dV_z for simplicity.

The inner product

$$\int_\Omega f\overline{g}\, dV$$

of $L^2(\Omega)$ is denoted by (f,g) $(= (f,g)_\Omega)$, and the norm of it by $\|f\|$ $(= \|f\|_\Omega)$.

Set $A^2(\Omega) := A(\Omega) \cap L^2(\Omega)$, and call the elements of this set L^2 *holomorphic functions*. $A^2(\Omega)$ will be shown to be a closed subspace of $L^2(\Omega)$ in the present section.

For $k \in \mathbb{Z}_+ \cup \{\infty\} \cup \{\omega\}$, let $C^k(\Omega)$ be the set of all complex valued functions of class C^k on Ω, and $C_0^k(\Omega)$ the set of those whose supports are compact. It is clear that $C_0^\infty(\Omega) \subset L^2(\Omega)$. Let us recall that $L^2(\Omega)$ is a separable space and that $C_0^\infty(\Omega)$ is dense in $L^2(\Omega)$.

Let $L^2_{\mathrm{loc}}(\Omega)$ be the set of all *locally square integrable functions* on Ω, that is, complex valued measurable functions on Ω that are square integrable on compact subsets of Ω.

Evidently, $A(\Omega) \subset C^\infty(\Omega) \subset L^2_{\mathrm{loc}}(\Omega)$.

In general, let V' be the set of all complex linear functionals on a vector space V over $\mathbb{C}$.

If we define an element $\iota(f)$ in $C_0^\infty(\Omega)'$ for $f \in L^2_{\mathrm{loc}}(\Omega)$ by

$$\iota(f)(g) := (f, \overline{g}) \text{ for } g \in C_0^\infty(\Omega),$$

then ι is an injection from $L^2_{\mathrm{loc}}(\Omega)$ to $C_0^\infty(\Omega)'$, since $C_0^\infty(\Omega)$ is dense in $L^2_{\mathrm{loc}}(\Omega)$ with respect to the topology induced by the L^2 convergence on compact sets. Hence, $L^2_{\mathrm{loc}}(\Omega)$ is identified with a subset of $C_0^\infty(\Omega)'$ under ι.

For an element u in $C_0^\infty(\Omega)'$, define an element $\left(\frac{\partial}{\partial x}\right)^\beta u$ in $C_0^\infty(\Omega)'$ by

$$\left(\frac{\partial}{\partial x}\right)^\beta u(g) := (-1)^{\langle\beta\rangle} u\left(\left(\frac{\partial}{\partial x}\right)^\beta g\right).$$

The definition of $\left(\frac{\partial}{\partial x}\right)^\beta u$ is a natural generalization of the concept of the usual derivative, since $\iota\left(\left(\frac{\partial}{\partial x}\right)^\beta h\right) = \left(\frac{\partial}{\partial x}\right)^\beta \iota(h)$ for an element h in $C^\infty(\Omega)$, by integration by parts.

The following theorem provides a characterization of holomorphic functions.

THEOREM 1.10 (Theorem of L^2 Holomorphy).

$$A(\Omega) = \left\{ f \in L^2_{\text{loc}}(\Omega) \;\middle|\; \frac{\partial f}{\partial \overline{z}_j} = 0 \textit{ for } j = 1, \cdots, n \right\}.$$

Let us review, below, complex differential forms in order to prepare some calculations that are used in the proof of this theorem. A *differential form of degree* r (or r*-form*) on Ω $(\subset \mathbb{C})$ can be written as

$$\sum_{I,J} u_{IJ}\, dz_I \wedge d\overline{z}_J, \text{ where } u_{IJ} = \operatorname{sgn}\begin{pmatrix} I \\ I' \end{pmatrix} \operatorname{sgn}\begin{pmatrix} J \\ J' \end{pmatrix} u_{I'J'}$$

for multi-indices $I = (i_1, \ldots, i_k)$ and $J = (j_1, \ldots, j_l)$ with $k + l = r$ whose components are taken from the natural numbers from 1 to n, and

$$\begin{aligned} dz_I &= dz_{i_1} \wedge \ldots \wedge dz_{i_k}, \\ d\overline{z}_J &= d\overline{z}_{j_1} \wedge \ldots \wedge d\overline{z}_{j_l}. \end{aligned}$$

A differential form is often written as

$$u = {\sum_{I,J}}' u_{IJ}\, dz_I \wedge d\overline{z}_J,$$

in which the multi-indices I and J are only allowed to have strictly increasing components. In this case, u is said to be a *differential form of type* (p, q), or simply a (p, q)*-form*, if the lengths k and l of I and J are equal to the given constants p and q, respectively.

Let $C^{(r)}(\Omega)$ be the set of all r-forms of class C^∞ on Ω, and let $C^{p,q}(\Omega)$ be the set of all (p, q)-forms of class C^∞ on Ω. In general, let $C^\infty(K)$ be the set of all complex valued functions of class C^∞ on a subset K of $\mathbb{C}^n$. When $K = \overline{\Omega}$, we define $C^{p,q}(\overline{\Omega}) \subset C^{p,q}(\Omega)$ and $C^{(r)}(\overline{\Omega}) \subset C^{(r)}(\Omega)$ in a similar fashion.

Let us give an example of a calculation in which differential forms are effectively used.

Given a holomorphic mapping $F = (f_1, \cdots, f_n) : \Omega \to \mathbb{C}^n$, we have

$$\begin{aligned}&F^*(dz_1 \wedge \ldots \wedge dz_n \wedge d\overline{z}_1 \wedge \ldots \wedge d\overline{z}_n)\\ &= \left| \det \left(\frac{\partial f_j}{\partial z_k} \right)_{j,k} \right|^2 dz_1 \wedge \ldots \wedge dz_n \wedge d\overline{z}_1 \wedge \ldots \wedge d\overline{z}_n.\end{aligned}$$

From this, the Jacobian of F with respect to the real coordinates $(x_1, x_2, \cdots, x_{2n})$ is equal to $\left| \det \left(\frac{\partial f_j}{\partial z_k} \right)_{j,k} \right|^2$.

The *complex exterior derivative operators*

$$\partial : C^{p,q}(\Omega) \to C^{p+1,q}(\Omega)$$

of type $(1,0)$ and

$$\overline{\partial} : C^{p,q}(\Omega) \to C^{p,q+1}(\Omega)$$

of type $(0,1)$ are defined respectively by

$$\begin{aligned}\partial\Big({\sum_{I,J}}' u_{IJ} dz_I \wedge d\overline{z}_J\Big) &= {\sum_{I,J}}' \sum_k \frac{\partial u_{IJ}}{\partial z_k} dz_k \wedge dz_I \wedge d\overline{z}_J,\\ \overline{\partial}\Big({\sum_{I,J}}' u_{IJ} dz_I \wedge d\overline{z}_J\Big) &= {\sum_{I,J}}' \sum_k \frac{\partial u_{IJ}}{\partial \overline{z}_k} d\overline{z}_k \wedge dz_I \wedge d\overline{z}_J.\end{aligned}$$

From this definition, it is obvious that the ordinary exterior derivative d is equal to $\partial + \overline{\partial}$.

Let $L^{p,q}_{\mathrm{loc}}(\Omega)$ denote the set of all differential forms of type (p,q) whose coefficients u_{IJ} are elements in $L^2_{\mathrm{loc}}(\Omega)$, and let $L^{p,q}(\Omega)$ be the set of those whose coefficients are in $L^2(\Omega)$. $L^{(r)}_{\mathrm{loc}}(\Omega)$ and $L^{(r)}(\Omega)$ are defined similarly. Also, $C^{p,q}_0(\Omega)$ expresses the subset of $C^{p,q}(\Omega)$ whose elements have compact supports, and $C^{(r)}_0(\Omega)$ the subset of $C^{(r)}(\Omega)$ whose elements have compact supports.

Using this notation, through the involution

$$\begin{array}{ccc} L^{(r)}_{\mathrm{loc}}(\Omega) \times C^{(2n-r)}_0(\Omega) & \longrightarrow & \mathbb{C} \\ \Downarrow & & \Downarrow \\ (u, v) & \longmapsto & \displaystyle\int_\Omega u \wedge v, \end{array}$$

$L^{(r)}_{\mathrm{loc}}(\Omega)$ is identified with a subspace of $C_0^{(2n-r)}(\Omega)'$, and similarly $L^{p,q}_{\mathrm{loc}}(\Omega)$ with a subspace of $C_0^{n-p,n-q}(\Omega)'$. Accordingly, the domains of definition for the exterior derivative operators d, ∂, and $\overline{\partial}$ extend to $L^{(r)}_{\mathrm{loc}}(\Omega)$ or $L^{p,q}_{\mathrm{loc}}(\Omega)$.

The proof of Theorem 1.10 needs the mean-value property for differentiable solutions of the Cauchy-Riemann equations.

PROPOSITION 1.11. *Assume that $f \in C^1(\Omega)$ and $\overline{\partial} f = 0$.*

1. *For an arbitrary n-open ball $\mathbb{B}(a, R) \Subset \Omega$, it follows that*

$$\frac{1}{\mathrm{Vol}(\partial\mathbb{B}(a,R))} \int_{\partial\mathbb{B}(a,\,R)} f\, dS = f(a), \tag{1.4}$$

where dS denotes the volume element of $\partial\mathbb{B}(a,R)$ induced by the Euclidean metric, and we set

$$\mathrm{Vol}(\partial\mathbb{B}(a,R)) := \int_{\partial\mathbb{B}(a,\,R)} dS = \frac{2\pi^n R^{2n-1}}{(n-1)!}.$$

2. $f \in A(\Omega)$.

PROOF OF (1). Set $g(z) := f(Rz+a) - f(a)$; then (1.4) is equivalent to

$$\int_{\partial\mathbb{B}^n} g\, dS = 0. \tag{1.5}$$

Now that the restriction of a $(2n-1)$-form $\partial \log|z| \wedge \left(\bigwedge^{n-1} \partial\overline{\partial} \log|z|\right)$ to $\partial\mathbb{B}^n$ is not equal to 0 but unitarily invariant, (1.5) is equivalent to

$$\int_{\partial\mathbb{B}^n} g(z)\, \partial \log|z| \wedge \left(\bigwedge^{n-1} \partial\overline{\partial} \log|z|\right) = 0.$$

In general, for a function u of class C^1 on $\overline{\mathbb{B}^n}$, Stokes' formula implies

$$\begin{aligned} &\int_{\partial\mathbb{B}^n} u(z)\, \partial \log|z| \wedge \left(\bigwedge^{n-1} \partial\overline{\partial} \log|z|\right) \\ &\quad = \int_{\mathbb{B}^n \setminus \mathbb{B}^n(0,\,\varepsilon)} d\left\{u(z)\, \partial \log|z| \wedge \left(\bigwedge^{n-1} \partial\overline{\partial} \log|z|\right)\right\} \\ &\qquad + \int_{\partial\mathbb{B}^n(0,\,\varepsilon)} u(z)\, \partial \log|z| \wedge \left(\bigwedge^{n-1} \partial\overline{\partial} \log|z|\right), \end{aligned} \tag{1.6}$$

for $0 < \varepsilon < 1$. In the case $u = g$, since the first term of the right hand side is 0 by the condition $\bar\partial g = 0$, it follows that

$$\int_{\partial \mathbb{B}^n} g(z)\, \partial \log |z| \wedge \left(\bigwedge^{n-1} \partial\bar\partial \log |z| \right)$$
$$= \int_{\partial \mathbb{B}^n(0,\, \varepsilon)} g(z)\, \partial \log |z| \wedge \left(\bigwedge^{n-1} \partial\bar\partial \log |z| \right).$$

Therefore, we obtain (1.5) by letting $\varepsilon \searrow 0$, since $g(0) = 0$. □

REMARK. When $n = 1$, in the above argument, take $g(z) = 1$, and replace z by an element f in $C^1(\overline{\Delta}) \cap \operatorname{Ker} \bar\partial$ that has no zero point on $\partial\Delta$. A similar calculation provides

Argument Principle:

$$\int_{\partial\Delta} \partial \log |f(z)| = \pi\sqrt{-1}\, n_f,$$

where n_f denotes the sum of the orders of zeros of f in Δ.

Before proceeding to the proof of (2), we need to prepare the following proposition:

PROPOSITION 1.12. *Under the same assumption as in Proposition 1.11,*

$$\frac{1}{\operatorname{Vol}(\mathbb{B}(a, R))} \int_{\mathbb{B}(a,\, R)} f\, dV = f(a), \tag{1.7}$$

where

$$\operatorname{Vol}(\mathbb{B}(a, R)) = \frac{\pi^n R^{2n}}{n!}.$$

PROOF. This is due to (1.4) and Fubini's theorem. □

Apply the Cauchy-Schwarz inequality to the left hand side of (1.7); then we obtain

$$|f(a)|^2 \leqq \frac{1}{\operatorname{Vol}(\mathbb{B}(a, R))} \int_{\mathbb{B}(a,\, R)} |f|^2\, dV. \tag{1.8}$$

This inequality is called *Cauchy's estimate.* In (1.8), the condition for $\mathbb{B}(a, R)$ may be relaxed to $\mathbb{B}(a, R) \subset \Omega$. (The right hand side is allowed to be ∞.)

Combine Cauchy's estimate with Theorem 1.6; then it turns out that $A(\Omega)$ is a closed subspace of $L^2_{\mathrm{loc}}(\Omega)$ with respect to the topology induced by the L^2 convergence on compact sets. From this, the separability of $A(\Omega)$ and $A^2(\Omega)$ follows.

PROOF OF (2). If f were of class C^1 on Ω, $\bar\partial f = 0$, and $f \notin A(\Omega)$, then there would exist $\mathbb{B}(a, R) \Subset \Omega$ such that the orthogonal projection $P : L^2(\mathbb{B}(a, R)) \to A^2(\mathbb{B}(a, R))$ does not map $u := f|\mathbb{B}(a, R)$ to itself. Hence, $g := Pu - u$ satisfies $g \neq 0$, $\bar\partial g = 0$, and $g \perp A^2(\mathbb{B}(a, R))$.

However, if we fix an arbitrary element σ in $\operatorname{Aut}\mathbb{B}(a, R)$, then

$$0 = \int_{\mathbb{B}(a,\,R)} g(\zeta)\overline{h(\zeta)}\, dV_\zeta = \int_{\mathbb{B}(a,\,R)} g(\sigma(z))\overline{h(\sigma(z))} \left| \det\left(\frac{\partial \sigma_j}{\partial z_k}\right)\right|^2 dV_z$$

for any $h \in A^2(\mathbb{B}(a, R))$. Hence,

$$\det\left(\frac{\partial \sigma_j}{\partial z_k}\right)^{\pm} \in A(\mathbb{B}(a, R)) \cap C^\infty(\overline{\mathbb{B}(a, R)})$$

implies

$$g(\sigma(z)) \det\left(\frac{\partial \sigma_j}{\partial z_k}\right) \perp A^2(\mathbb{B}(a, R)).$$

Since clearly

$$\bar\partial\left(g(\sigma(z)) \det\left(\frac{\partial \sigma_j}{\partial z_k}\right)\right) = 0,$$

it follows from the mean-value property that

$$\begin{aligned} g(\sigma(a)) \det\left(\frac{\partial \sigma_j}{\partial z_k}\right)(a) &= \frac{1}{\operatorname{Vol}(\mathbb{B}(a, R))} \int_{\mathbb{B}(a,\,R)} g(\sigma(z)) \det\left(\frac{\partial \sigma_j}{\partial z_k}\right) dV_z \\ &= 0 \quad (\because 1 \in A^2(\mathbb{B}(a, R))). \end{aligned}$$

Now that $\operatorname{Aut}\mathbb{B}(a, R)$ is transitive, it follows that $g \equiv 0$. This contradicts that $Pu \neq u$.

Therefore, from the assumption that $f \in C^1(\Omega)$ and $\bar\partial f = 0$, it must follow that $f \in A(\Omega)$. □

Let us review some fundamental facts on the regularization of elements in $L^2_{\text{loc}}(\Omega)$ before getting into the proof of Theorem 1.10.

Take a monotone decreasing function (in the broad sense) $\mu : \mathbb{R} \to [0, 1]$ of class C^∞ with $\operatorname{supp}\mu \subset (-\infty, 1)$ and

$$\int_0^\infty \mu(t) t^{2n-1}\, dt = 1,$$

and let

$$\mu_\varepsilon(z) := \frac{1}{\varepsilon^{2n}\mathrm{Vol}(\partial\mathbb{B}^n)}\mu\left(\frac{|z|}{\varepsilon}\right). \tag{1.9}$$

The main properties that μ_ε possesses are $\mu_\varepsilon \in C^\infty(\mathbb{C}^n)$, $\mu_\varepsilon \geqq 0$, $\operatorname{supp}\mu_\varepsilon \subset \mathbb{B}^n(0,\varepsilon)$,

$$\int_{\mathbb{C}^n} \mu_\varepsilon\, dV = 1,$$

and that μ_ε depends only on $|z|$ as a function. The monotonicity of μ will be convenient for later use.

Also, set

$$\Omega_\varepsilon := \left\{ z \in \Omega \;\middle|\; \inf_{w\in\partial\Omega} |z-w| > \varepsilon \right\}$$

for a positive number ε.

If for an element f in $L^2_{\mathrm{loc}}(\Omega)$, we put

$$f_\varepsilon(z) := \int_{\mathbb{C}^n} f(z+\zeta)\mu_\varepsilon(\zeta)\, dV_\zeta\,,$$

then $f_\varepsilon \in C^\infty(\Omega_\varepsilon)$ and f_ε *converges* to f *with respect to the* L^2 *norm on compact sets.* That is to say, for any relatively compact open subset Ω' of Ω,

$$\lim_{\varepsilon\to 0} \|f_\varepsilon - f\|_{\Omega'} = 0. \tag{1.10}$$

(For the proof, see [**28**] for instance.)

f_ε is called the ε*-regularization* of f. Later, this terminology will be used for differential forms, with the same meaning.

PROOF OF THEOREM 1.10. If $f \in L^2_{\mathrm{loc}}(\Omega)$ and $\overline{\partial} f = 0$, then

$$\frac{\partial f_\varepsilon(z)}{\partial \overline{z}_j} = \int_{\mathbb{C}^n} f(\zeta)\frac{\partial}{\partial \overline{z}_j}\mu_\varepsilon(\zeta - z)\, dV_\zeta = -\int_{\mathbb{C}^n} f(\zeta)\frac{\partial}{\partial \overline{\zeta}_j}\mu_\varepsilon(\zeta - z)\, dV_\zeta = 0.$$

Hence, what has been previously shown implies $f_\varepsilon \in A(\Omega_\varepsilon)$. Therefore, the mean-value property becomes applicable to f_ε and results in

$$(f_\varepsilon)_\delta(z) = \int_{\mathbb{C}^n} f_\varepsilon(z+\zeta)\mu_\delta(\zeta)\, dV_\zeta = f_\varepsilon(z) \text{ for } z \in \Omega_{\varepsilon+\delta}.$$

On the other hand, the right hand side of these equations is equal to

$$\begin{aligned}&\int_{\mathbb{C}^n}\left(\int_{\mathbb{C}^n} f(z+\zeta+\xi)\mu_\varepsilon(\xi)\,dV_\xi\right)\mu_\delta(\zeta)\,dV_\zeta\\&=\int_{\mathbb{C}^n}\left(\int_{\mathbb{C}^n} f(z+\xi+\zeta)\mu_\delta(\zeta)\,dV_\zeta\right)\mu_\varepsilon(\xi)\,dV_\xi\\&=\int_{\mathbb{C}^n} f_\delta(z+\zeta)\mu_\varepsilon(\zeta)\,dV_\zeta=(f_\delta)_\varepsilon(z).\end{aligned}$$

Therefore, $f_\varepsilon = f_\delta$ on $\Omega_{\varepsilon+\delta}$. Additionally, if (1.10) is taken into account, then $f_\varepsilon = f$ on Ω_ε. This proves $f \in A(\Omega)$, since we have shown that $f_\varepsilon \in A(\Omega_\varepsilon)$. □

As an application of Theorem 1.10, we obtain a continuation theorem that describes a sufficient condition, in terms of the Lebesgue measure, for a closed subset E of Ω to satisfy $A^2(\Omega \setminus E) = A^2(\Omega)$. Below, let $m(B)$ denote the Lebesgue measure of B.

THEOREM 1.13. *Assume that for an arbitrary point z_0 in a closed subset E of Ω, there exists a neighborhood U of z_0 in Ω such that*

$$\liminf_{\varepsilon\to 0}\varepsilon^{-2}m\left(\left\{z\in U \;\middle|\; \inf_{w\in E}|z-w|<\varepsilon\right\}\right)<\infty. \tag{1.11}$$

Then $A^2(\Omega \setminus E) = A^2(\Omega)$.

PROOF. Set $d_E(z) := \inf_{w\in E}|z-w|$. Also, take a C^∞ function $\rho : \mathbb{R} \to [0,1]$ such that $\rho\Big|\left(-\infty, \frac{1}{2}\right) = 1$ and $\rho|(1,\infty) = 0$, and define a function χ_ε on Ω by

$$\chi_\varepsilon(z) := \rho\left(\frac{d_E(z)}{\varepsilon}\right), \text{ where } \varepsilon > 0.$$

$d_E(z)$ is almost everywhere differentiable, since it is Lipschitz continuous. Accordingly, so is χ_ε, and

$$\left|\frac{\partial\chi_\varepsilon(z)}{\partial x_j}\right| \leqq \frac{1}{\varepsilon}\sup_{t\in\mathbb{R}}|\rho'(t)|$$

almost everywhere on Ω.

Suppose that $f \in A^2(\Omega \setminus E)$. Since the given condition implies $L^2(\Omega \setminus E) = L^2(\Omega)$, it suffices to show that for any element u in $C_0^\infty(\Omega)$,

$$\int_\Omega f\frac{\partial u}{\partial \overline{z}_j}\,dV = 0 \text{ for } j = 1, \cdots, n.$$

For this purpose, divide the left hand side of the above equation into

$$\int_\Omega f\frac{\partial}{\partial\overline{z}_j}(\chi_\varepsilon u)\,dV + \int_\Omega f\frac{\partial}{\partial\overline{z}_j}((1-\chi_\varepsilon)u)\,dV.$$

First, since $\overline{\partial}f = 0$ on $\Omega\setminus E$, integration by parts implies

$$\int_\Omega f\frac{\partial}{\partial\overline{z}_j}((1-\chi_\varepsilon)u)\,dV = 0. \tag{1.12}$$

On the other hand,

$$\left|\int_\Omega f\frac{\partial}{\partial\overline{z}_j}(\chi_\varepsilon u)\,dV\right| \leqq \left|\int_\Omega f\frac{\partial\chi_\varepsilon}{\partial\overline{z}_j}u\,dV\right| + \left|\int_\Omega f\chi_\varepsilon\frac{\partial u}{\partial\overline{z}_j}\,dV\right|.$$

The first term on the right hand side satisfies

$$\left|\int_\Omega f\frac{\partial\chi_\varepsilon}{\partial\overline{z}_j}u\,dV\right|^2 \leqq \varepsilon^{-2}\sup|u|^2\sup|\rho'|^2\cdot m(E_{u,\varepsilon})\int_{E_{u,\varepsilon}}|f|^2\,dV\,,$$

where we set

$$E_{u,\varepsilon} := \{z \mid d_E(z) < \varepsilon\}\cap\operatorname{supp}u.$$

From the assumption, the inferior limit of the right hand side equals 0 as $\varepsilon\to 0$. Moreover, the second term satisfies

$$\left|\int_\Omega f\chi_\varepsilon\frac{\partial u}{\partial\overline{z}_j}\,dV\right|^2 \leqq \sup\left|\frac{\partial u}{\partial\overline{z}_j}\right|^2\int_{E_{u,\varepsilon}}|f|^2\,dV \longrightarrow 0.$$

Hence, by combining these results we obtain

$$\liminf_{\varepsilon\to 0}\left|\int_\Omega f\frac{\partial}{\partial\overline{z}_j}(\chi_\varepsilon u)\,dV\right| = 0. \tag{1.13}$$

Now (1.12) and (1.13) imply the desired conclusion. □

For a holomorphic function f on Ω, the zero set $f^{-1}(0)$ of f is denoted by $V(f)$ or $V(f(z))$. An important example of applying Theorem 1.13 is given in the case $E = V(f)$ as follows (though this result will not be used in this book):

PROPOSITION 1.14. *If $V(f)$ does not contain an interior point, then $A^2(\Omega\setminus V(f)) = A^2(\Omega)$.*

SKETCH OF THE PROOF. If for a point a in $V(f)$,

$$f(z) = \sum_{\langle\alpha\rangle\geqq m} c_\alpha(z-a)^\alpha,\ \text{ where } c_\alpha\neq 0 \text{ for some } \langle\alpha\rangle = m,$$

on some neighborhood of a in Ω, then by applying an appropriate coordinate transformation

$$z - a = Bw \text{ for } w \in \mathbb{C}^n \text{ and a complex } n \times n \text{ regular matrix } B,$$

we obtain

$$f(Bw + a) = w_n^m g(w) + c_1(w')w_n^{m-1} + \cdots + c_m(w'),$$

where $w' := (w_1, \cdots, w_{n-1})$, $c_1(0) = \cdots = c_m(0) = 0$, and $g(0) \neq 0$.

Therefore, by restricting the projection $w \mapsto w'$ to a neighborhood of 0, the intersection of the preimage of each point with $V(f(Bw + a))$ consists of at most m points (by the argument principle). Hence, from Fubini's theorem it turns out that for an ε-neighborhood $V(f)_\varepsilon$ of $V(f)$ and a relatively compact subset U of Ω, the Lebesgue measure of $V(f)_\varepsilon \cap U$ is evaluated to be the infinitesimal of order 2 with respect to ε. This means that $V(f)$ satisfies the condition of Theorem 1.13. □

REMARK. We describe two facts that are related to Theorem 1.13.

1. In the case $n = 1$, a necessary and sufficient condition for $A^2(\Omega \setminus E) = A^2(\Omega)$ is known. (We refer the reader to Theorem 5.13 in § 5.4.)
2. Shiffman [**41**] has shown that $A(\Omega \setminus E) = A(\Omega)$ in the special case when the left hand side of (1.11) is equal to 0.

1.3. Reinhardt Domains

It is fundamental that the convergence range of the Taylor series at the origin for a holomorphic function defined on Δ^n is a set containing Δ^n. In general, however, the convergence range of a power series in several variables can take various forms other than Δ^n. In this section, we will mention general properties that such sets possess.

Let Ω be a domain in $\mathbb{C}^n$, namely, a connected open set.

DEFINITION 1.15. Ω is said to be a *Reinhardt domain* with center a if

$$(a_1 + \zeta_1 \cdot (z_1 - a_1), \cdots, a_n + \zeta_n \cdot (z_n - a_n)) \in \Omega \tag{1.14}$$

for any $z \in \Omega$ and any $\zeta \in (\partial\Delta)^n$. Also, Ω is called a *complete Reinhardt domain* with center a if (1.14) holds for any $z \in \Omega$ and any $\zeta \in \Delta^n$.

Polydiscs and complex open balls are examples of complete Reinhardt domains. Clearly, a complete Reinhardt domain contains the center in it.

Let D be a complete Reinhardt domain. Assume that the center of D is 0 for simplicity. The next proposition follows immediately from Proposition 1.4.

PROPOSITION 1.16. *For a holomorphic function f on a complete Reinhardt domain D with center* 0, *the power series*

$$P(0, f) = \sum_{\alpha} \frac{f^{(\alpha)}(0)}{\alpha!} z^{\alpha}$$

converges on D.

Define the *logarithmic image* $\log D$ of D by

$$\log D := \{x \in (\mathbb{R} \cup \{-\infty\})^n \mid e^x := (e^{x_1}, \cdots, e^{x_n}) \in D\}.$$

Let $(\log D)^\wedge$ be the convex hull of $\log D$, and set

$$\widehat{D} := \{z \in \mathbb{C}^n \mid (\log |z_1|, \cdots, \log |z_n|) \in (\log D)^\wedge\}.$$

THEOREM 1.17. *For $f \in A(D)$, $P(0, f)$ converges on $\widehat{D}$.*

PROOF. Take any $\zeta \in \widehat{D}$, and set $r := |\zeta| := (|\zeta_1|, \cdots, |\zeta_n|)$. Then, from the definition of $\widehat{D}$, r can be written as

$$r = (r_1'^{\,t} r_1''^{\,1-t}, \cdots, r_n'^{\,t} r_n''^{\,1-t}) =: r'^{t} r''^{1-t}$$

for some two points r' and r'' in $D \cap [0, \infty)^n$ and some $0 \leqq t \leqq 1$. Let c_α be the coefficient of z^α in $P(0, f)$. Since

$$\sum_{\alpha} |c_\alpha| r'^{\alpha} < \infty \text{ and } \sum_{\alpha} |c_\alpha| r''^{\alpha} < \infty,$$

there exists a constant M such that

$$|c_\alpha| r'^{\alpha} < M \text{ and } |c_\alpha| r''^{\alpha} < M$$

for any α. Hence, it follows that

$$\begin{aligned} |c_\alpha| r^\alpha &= |c_\alpha| (r'^{t} r''^{1-t})^\alpha \\ &= (|c_\alpha| r'^{\alpha})^t (|c_\alpha| r''^{\alpha})^{1-t} < M^t \cdot M^{1-t} = M. \end{aligned}$$

This implies the convergence of $P(0, f)$ on $\Delta(0, r)$. Moreover, noting that

$$\widehat{D} = \bigcup_{\zeta \in \widehat{D}} \Delta(0, |\zeta|),$$

we conclude that the convergence of $P(0, f)$ holds on $\widehat{D}$ as well. □

A Reinhardt domain whose logarithmic image is convex is said to be *logarithmically convex*. Let us show one property of logarithmically convex complete Reinhardt domains.

PROPOSITION 1.18. *For a logarithmically convex complete Reinhardt domain D with the origin centered and an exterior point a of D, there exists a monomial $m_a(z)$ such that*

$$\sup_{z\in D} |m_a(z)| < m_a(a) = 1.$$

PROOF. Since $\log D$ is convex, there are $\gamma \in \mathbb{Z}^n$ and $\delta \in \mathbb{R}$ such that

$$\begin{cases} \sup\{\langle x, \gamma\rangle + \delta \mid x \in \log D\} < 0, \\ e^{\delta}|a^{\gamma}| = 1\,. \end{cases}$$

Hence, it suffices to put $m_a(z) = e^{\delta}|a^{\gamma}|a^{-\gamma} \cdot z^{\gamma}$. □

COROLLARY 1.19. *Let D be the same as above. For an arbitrary point a of ∂D, there exists an element F in $A(D)$ that satisfies*

$$\varlimsup_{z\to a} |F(z)| = \infty.$$

PROOF. Take an increasing sequence D_j of relatively compact subdomains of D that are logarithmically convex complete Reinhardt domains. Take a sequence a_j of points in D that converges to a and satisfies $a_j \in D_{j+1} \backslash D_j$. Let $m_{a_j}(z)$ be such a monomial as above that is determined for each domain D_j and point a_j. Choose sequences $c_j \in \mathbb{R}$ and $\nu_j \in \mathbb{N}$ such that

$$\begin{cases} c_k = \left|\sum_{j=1}^{k-1} c_j m_{a_j}(a_k)^{\nu_j}\right| + k, \\ \sup_{z\in D_k} |c_k m_{a_k}(z)^{\nu_k}| < \dfrac{1}{2^k}\,. \end{cases} \tag{1.15}$$

Then the series

$$\sum_{j=1}^{\infty} c_j m_{a_j}(z)^{\nu_j}$$

converges uniformly on compact sets in D. Denoting the sum of the series by $F(z)$, Theorem 1.6 implies $F \in A(D)$. On the other hand, from (1.15), we clearly have $|F(a_k)| > k - 1$ for each k. This results in $\varlimsup_{z\to a} |F(z)| = \infty$. □

The above proof shows that the conclusion of Corollary 1.19 may be strengthened as follows:

> *"For any sequence $\{a_j\}$ of points in D that has no accumulating point in D, there exists an element F in $A(D)$ such that*
>
> $$\overline{\lim_{j\to\infty}}\, |F(a_j)| = \infty\,."$$

General open sets that possess this property will be discussed later. It seems interesting, at this point, to study the convergence range of a series $\sum_{k=0}^{\infty} f_k$ whose terms are homogeneous polynomials f_k of degree k as a slight generalization of a power series, but we do not do so in this book. H. Cartan wrote an article [**8**] about this problem, and we refer the reader to it.

CHAPTER 2

Rings of Holomorphic Functions and $\overline{\partial}$ Cohomology

From Weierstrass's theorem, it follows that $A(\Omega)$ is a complete topological ring with respect to the topology induced by uniform convergence on compact sets. Let us consider the structure of the *spectrum* of $A(\Omega)$, i.e., the space of all maximal closed ideals of $A(\Omega)$ equipped with the weak topology. Ω can be regarded as a subset of the spectrum of $A(\Omega)$, since each point of Ω corresponds to a maximal closed ideal consisting of functions whose value is zero at that point. In what case does Ω coincide with the spectrum? If this is the case, from the Banach-Steinhaus theorem, it follows that for a discrete sequence $\{\xi_k\}_{k=1}^{\infty}$ of points in Ω, there always exists an element f in $A(\Omega)$ such that $\overline{\lim\limits_{k\to\infty}} |f(\xi_k)| = \infty$; does there exist any f that satisfies $f(\xi_k) = k$ $(k = 1, 2, \cdots)$, for instance?

By replacing these problems with those of solving the Cauchy-Riemann equations of inhomogeneous form, let us connect the spectrum, a concept of topological algebra, with $\overline{\partial}$ cohomology, an analytic concept.

2.1. Spectra and the $\overline{\partial}$ Equation

The spectrum of $A(\Omega)$ is denoted by $\operatorname{Spec}_m A(\Omega)$.

PROPOSITION 2.1. *If for an arbitrary sequence $\{f_k\}_{k=1}^{\infty} \subset A(\Omega)$ of functions that has no common zero point, there exists a sequence $\{g_k\}_{k=1}^{\infty}$ of functions in $A(\Omega)$ such that*

$$\sum_{k=1}^{\infty} f_k g_k = 1, \tag{2.1}$$

then $\Omega = \operatorname{Spec}_m A(\Omega)$.

PROOF. This is clear from the paracompactness of Ω and the definition of $\operatorname{Spec}_m A(\Omega)$. □

The converse of Proposition 2.1 may seem self-evident, but in fact it is not.[1] The proof of the converse requires a good amount of preparation, and is deferred until Chapter 5.

Let us characterize a sequence $\{g_k\}_{k=1}^{\infty}$ of functions that satisfies (2.1) as a solution of the $\overline{\partial}$ equation with some restraints.

We can assume that the series

$$h = \sum_{k=1}^{\infty} |f_k|^2$$

converges uniformly on compact sets by replacing the given sequence of functions f_k with $\varepsilon_k f_k$ ($\varepsilon_k \neq 0$) if necessary. The estimate (1.2) and the Ascoli-Arzelà theorem imply $h \in C^{\infty}(\Omega)$, and h has no zero point by assumption. Hence,

$$h_k := \frac{\overline{f_k}}{h} \in C^{\infty}(\Omega),$$

and

$$\sum_{k=1}^{\infty} f_k h_k = 1 \text{ uniformly on compact sets.}$$

Therefore,

$$\sum_{k=1}^{\infty} f_k \overline{\partial} h_k = 0.$$

The uniform convergence of the left hand side on compact sets is due to the same reason that $h \in C^{\infty}(\Omega)$.

If there exists a sequence $\{u_k\}_{k=1}^{\infty}$ of functions in $C^{\infty}(\Omega)$ such that

$$(2.2) \qquad \begin{cases} \sum_{k=1}^{\infty} f_k u_k = 0 \text{ uniformly on compact sets, and} \\ \overline{\partial} u_k = \overline{\partial} h_k \text{ for } k = 1, 2, \cdots, \end{cases}$$

then by setting $g_k = h_k - u_k$, we obtain a sequence of functions that meets the condition (2.1). Conversely, given g_k, the functions $u_k := h_k - g_k$ provide a solution of (2.2). Therefore, the equation (2.1) for the system of unknown functions g_k is equivalent to the equations in (2.2) for the system of unknown functions u_k.

As the second problem, we consider a condition for the restriction map $A(\Omega) \to \mathbb{C}^{\Gamma}$ to be surjective for a given set Γ of points that has no accumulating point inside Ω. Let us convert this simplest *interpolation problem* into one for the $\overline{\partial}$ equation.

---

[1] Because $\sum_{k=1}^{\infty} f_k g_{j,k} \to 1 \, (j \to \infty)$ does not necessarily imply the convergence of $\{g_{j,k}\}_{j=1}^{\infty}$.

Fix a system $\{U_\xi\}_{\xi\in\Gamma}$ of mutually disjoint open sets with $\xi \in U_\xi \subset \Omega$, and take a function ρ of class C^∞ on Ω such that $\operatorname{supp}\rho \subset \bigcup_{\xi\in\Gamma} U_\xi$ and $\rho \equiv 1$ on some neighborhood of Γ. For an arbitrary $b \in \mathbb{C}^\Gamma$, define $\widetilde{b} \in C^\infty(\Omega)$ by

$$\widetilde{b}(z) = \begin{cases} b(\xi)\rho(z) & \text{for } z \in U_\xi, \\ 0 & \text{for } z \in \left(\bigcup_{\xi\in\Gamma} U_\xi\right)^c. \end{cases}$$

Then $\widetilde{b}$ is a C^∞ extension of b to Ω with the property that $\overline{\partial}\,\widetilde{b} = 0$ on some neighborhood of Γ. Therefore, a necessary and sufficient condition for an element f in $A(\Omega)$ that satisfies $f \mid \Gamma = b$ to exist is that the $\overline{\partial}$ equation

$$\begin{cases} \overline{\partial} u = \overline{\partial}\,\widetilde{b}, \\ u \mid \Gamma = 0 \end{cases} \tag{2.3}$$

have a C^∞ solution u.

2.2. Extension Problems and the $\overline{\partial}$ Equation

The proper way to describe the extension problem of holomorphic functions should be more general, as seen below.

DEFINITION 2.2. A closed subset X of Ω is said to be an *analytic subset* if for any point x_0 of X, there exist a neighborhood U of x_0 in $\mathbb{C}^n$ and a system of functions $\{f_\alpha\}_{\alpha\in\Lambda} \subset A(U)$ (Λ may be an infinite set) such that

$$X \cap U = \{z \in U \mid f_\alpha(z) = 0 \text{ for } \alpha \in \Lambda\}.$$

We call $\{f_\alpha\}_{\alpha\in\Lambda}$ a *system of local defining functions* of X on U or simply around x_0.

A discrete subset and the intersection of Ω with a complex m-dimensional hyperplane are examples of analytic subsets.

For a function on an analytic subset X of Ω, the concept of holomorphic function is generalized as follows:

DEFINITION 2.3. A function f on X is *holomorphic* if for every point x_0 of X, there exist a neighborhood U of x_0 in $\mathbb{C}^n$ and an element F in $A(U)$ such that $F \mid U \cap X = f \mid U \cap X$.

By this definition, the interpolation problem can be grasped as the extension problem, to extend holomorphic functions defined on a 'lower-dimensional' subset inside Ω to functions on Ω. This understanding is geometrically richer and more interesting.

For the sake of brevity, we call a complex 1-dimensional hyperplane a *complex line* and a hyperplane of complex codimension 1 simply a *hyperplane.*

It is self-evident that when X is the intersection of a complex m-dimensional hyperplane with Ω, the above definition of holomorphic function coincides with the usual definition, in which X is regarded as an open set in $\mathbb{C}^m$.

Putting $L := \{z \in \mathbb{C}^n \mid z_1 = \cdots = z_{n-m} = 0\}$, let us convert the surjectivity problem of the restriction map

$$A(\Omega) \longrightarrow A(\Omega \cap L)$$

into the $\overline{\partial}$ equation.

Let $f = f(z_{n-m+1}, \cdots, z_n)$ be a holomorphic function on $\Omega \cap L$. Then a function $\widehat{f}(z) := f(0, \cdots, 0, z_{n-m+1}, \cdots, z_n)$ is holomorphic on an open set $W := \{z \in \mathbb{C}^n \mid (0, \cdots, 0, z_{n-m+1}, \cdots, z_n) \in \Omega \cap L\}$.

Therefore, by taking a C^∞ function $\rho_W : W \to [0, 1]$ that satisfies

$$\begin{cases} \operatorname{supp}(\rho_W - 1) \cap L = \emptyset, \\ \operatorname{supp} \rho_W \cap \partial\Omega = \emptyset, \end{cases}$$

and by letting $\widetilde{f}$ be the trivial extension of $\rho_W \widehat{f}$ to Ω, we see that $\widetilde{f} \in C^\infty(\Omega)$ and $\widetilde{f} \mid \Omega \cap L = f$.

Hence, the existence problem of a holomorphic extension of f to Ω can be replaced by the solvability problem of

$$\begin{cases} \overline{\partial} u = \overline{\partial} \widetilde{f}, \\ u \mid \Omega \cap L = 0 \end{cases} \tag{2.4}$$

in the same way as (2.3).

For a general analytic subset X, it is difficult to construct directly and precisely a local extension of f that corresponds to $\widehat{f}$ in the above argument. For this reason, careful consideration on a system of local defining functions is inevitable in order to discuss, from a general point of view, the extension problem of holomorphic functions without restricting ourselves to the case of discrete subsets (see § 5.1).

2.3. $\overline{\partial}$ Cohomology and Serre's Condition

Let L be as defined in § 2.2. Let us develop the argument to extend holomorphic functions on $\Omega \cap L$ to Ω. The result will relate to the structure problem of spectra.

DEFINITION 2.4. The $\overline{\partial}$ *cohomology group* of type (p, q) on Ω is, by definition,

$$H^{p,q}(\Omega) := \operatorname{Ker} \overline{\partial} \cap C^{p,q}(\Omega) / \operatorname{Im} \overline{\partial} \cap C^{p,q}(\Omega),$$

where the domain for the operator $\overline{\partial}$ is restricted to the space of C^∞ differential forms on Ω, and in general the kernel and image of a linear map T are denoted by $\operatorname{Ker} T$ and $\operatorname{Im} T$, respectively.[2]

From the above definition and the result in Chapter 1, it follows that $H^{0,0}(\Omega) = A(\Omega)$.

When there is an inclusion relation $\Omega_1 \supset \Omega_2$ between open sets Ω_1 and Ω_2, the restriction map $C^{p,q}(\Omega_1) \to C^{p,q}(\Omega_2)$ is well-defined and, being commutative with $\overline{\partial}$, induces a homomorphism from $H^{p,q}(\Omega_1)$ to $H^{p,q}(\Omega_2)$. This is called the *restriction homomorphism*.

On the one hand, the restriction map $C^{p,q}(\Omega) \to C^{p,q}(\Omega \cap L)$ is well-defined as the restriction of differential forms on Ω to the submanifold $\Omega \cap L$ and, again by the commutativity with $\overline{\partial}$, induces a homomorphism from $H^{p,q}(\Omega)$ to $H^{p,q}(\Omega \cap L)$, which is also called the *restriction homomorphism*.

Set $L_j := \{z \in \mathbb{C}^n \mid z_1 = \cdots = z_{n-j} = 0\}$ for $0 \leqq j \leqq n$. (Hence, $L = L_m$.)

THEOREM 2.5. *The restriction map*

$$A(\Omega) \longrightarrow A(\Omega \cap L)$$

is a surjection if $H^{0,q}(\Omega) = \{0\}$ *for all* q *with* $1 \leqq q \leqq n - m$.

This is called *Serre's criterion*.

For the proof, we need one lemma.

LEMMA 2.6. *The restriction homomorphism*

$$\alpha : H^{p,q}(\Omega) \longrightarrow H^{p,q}(\Omega \cap L_{n-1})$$

is a surjection if $H^{p,q+1}(\Omega) = \{0\}$. *In particular,* $H^{p,q}(\Omega \cap L_{n-1}) = \{0\}$ *if* $H^{p,q+k}(\Omega) = \{0\}$ *for all* $k = 0, 1$.

---

[2] Im overlaps the notation for the imaginary part of a complex number, but there should not be any confusion.

PROOF. Take $v \in C^{p,q}(\Omega \cap L_{n-1}) \cap \operatorname{Ker} \overline{\partial}$. As in the case of a holomorphic function, there is a C^∞ extension $\widetilde{v}$ of v to Ω such that $\overline{\partial}\widetilde{v} = 0$ on some neighborhood of $\Omega \cap L_{n-1}$. Since

$$\frac{\overline{\partial}\widetilde{v}}{z_1} \in C^{p,q+1}(\Omega) \cap \operatorname{Ker} \overline{\partial},$$

the assumption implies that there is an element $\widetilde{u}$ in $C^{p,q}(\Omega)$ such that

$$\overline{\partial}\widetilde{u} = \frac{\overline{\partial}\widetilde{v}}{z_1}.$$

Hence, it follows that $\widetilde{v} - z_1\widetilde{u} \in \operatorname{Ker} \overline{\partial}$ and $\widetilde{v} - z_1\widetilde{u} \mid \Omega \cap L_{n-1} = v$, which shows that α is surjective. □

PROOF OF THEOREM 2.5. By applying Lemma 2.6, the assumption implies

$$H^{0,1}(\Omega \cap L_j) = \{0\} \text{ for } m + 1 \leqq j \leqq n.$$

Hence, from the same lemma again, the restriction map

$$A(\Omega \cap L_{j+1}) \longrightarrow A(\Omega \cap L_j)$$

is a surjection for $j \geqq m$. Therefore, the restriction map $A(\Omega) \to A(\Omega \cap L)$ is also a surjection. □

For the closure $\overline{\Omega}$ of Ω, let $H^{p,q}(\overline{\Omega}) := \varinjlim_{U \supset \overline{\Omega}} H^{p,q}(U)$, where the inductive system $\{H^{p,q}(U)\}$ is regarded with respect to the restriction homomorphisms, and U runs in the fundamental system of open neighborhoods of $\overline{\Omega}$. For the sake of consistency in notation, we set $A(\overline{\Omega}) := H^{0,0}(\overline{\Omega})$.

From Serre's criterion, the next theorem follows immediately.

THEOREM 2.7. *The restriction map*

$$A(\overline{\Omega}) \longrightarrow A(\overline{\Omega} \cap L)$$

is a surjection if $H^{0,q}(\overline{\Omega}) = \{0\}$ *for the range of* $1 \leqq q \leqq n - m$.

If Ω is a polydisc, one can easily show that $H^{0,q}(\overline{\Omega}) = \{0\}$ for $q \geqq 1$ by an elementary method. Although the proof of this statement is included in many books, we will describe it in detail, considering the important role that this result will play when we generalize the theorem of L^2 holomorphy to the one for $\overline{\partial}$ cohomology.

We begin with the following lemma.

LEMMA 2.8. *For a bounded closed set K of $\mathbb{C}$,*

$$\varinjlim_{U\supset K} H^{0,1}(U) = \{0\}. \tag{2.5}$$

PROOF. Let U be a neighborhood of K, and $v \in C^{0,1}(U)$ $(= C^{0,1}(U) \cap \operatorname{Ker}\overline{\partial})$. It suffices to show that the $\overline{\partial}$ equation $\overline{\partial}u = v$ has a solution under the assumption that $v \in C_0^{0,1}(U)$, by multiplying v, if necessary, by a function whose value is 1 on a neighborhood of K and whose support is contained in U.

Let v denote also the coefficient function in the given $(0,1)$-form v. Set

$$u(z) = \frac{1}{2\pi\sqrt{-1}} \int_{\mathbb{C}} \frac{v(\xi + z)}{\xi} d\xi \wedge d\overline{\xi}. \tag{2.6}$$

Then it follows that $u \in C^\infty(\mathbb{C})$, and

$$\frac{\partial u}{\partial \overline{z}} = \frac{1}{2\pi\sqrt{-1}} \int_{\mathbb{C}} \frac{v_{\overline{z}}(\xi + z)}{\xi} d\xi \wedge d\overline{\xi}. \tag{2.7}$$

Stokes' formula implies

$$v(z) = \frac{1}{2\pi\sqrt{-1}} \int_{\mathbb{C}} \frac{v_{\overline{\xi}}(\xi + z)}{\xi} d\xi \wedge d\overline{\xi}. \tag{2.8}$$

Noting that $v_{\overline{\xi}}(\xi + z) = v_{\overline{z}}(\xi + z)$, from (2.7) and (2.8) we derive

$$\frac{\partial u}{\partial \overline{z}} = v.$$

□

When v contains a parameter w, if v is of class C^∞ or holomorphic with respect to w, so is the above solution u, as we clearly see from (2.6). This fact also will be used in the following argument.

THEOREM 2.9.

$$H^{0,q}(\overline{\Delta^n}) = \{0\} \textit{ for all } q \geqq 1.$$

PROOF. Take a neighborhood U of $\overline{\Delta^n}$ that provides a representative $v \in C^{0,q}(U) \cap \operatorname{Ker}\overline{\partial}$ of an arbitrary element in $H^{0,q}(\overline{\Delta^n})$ $(q \geqq 1)$. Set

$$v := \sum_{\substack{I \not\ni n \\ |I|=q-1}}{}' v'_I d\overline{z}_n \wedge d\overline{z}_I + \sum_{\substack{J \not\ni n \\ |J|=q}}{}' v''_J d\overline{z}_J$$

with $v'_I, v''_J \in C^\infty(U)$,

where $|I|$ denotes the length of the multi-index I, and $I \not\ni n$ means that I does not contain n.

From Lemma 2.8 and the succeeding remark, there is an element $u_I^{(1)}$ in $C^\infty(\mathbb{C}^n)$ satisfying

$$\frac{\partial u_I^{(1)}}{\partial \overline{z}_n} = v'_I \tag{2.9}$$

on some neighborhood of $\overline{\Delta^n}$. By setting

$$u^{(1)} := \sum_{\substack{I \not\ni n \\ |I|=q-1}}' u_I^{(1)} d\overline{z}_I,$$

from (2.9) v is transformed into the following form that does not contain $d\overline{z}_n$ any more:

$$v - \overline{\partial} u^{(1)} = \sum_{\substack{I \not\ni n,n-1 \\ |I|=q-1}}' w'_I d\overline{z}_{n-1} \wedge d\overline{z}_I + \sum_{\substack{J \not\ni n,n-1 \\ |J|=q}}' w''_J d\overline{z}_J,$$

where w'_I and w''_J on the right hand side are holomorphic with respect to z_n.

Therefore, we can take, this time, a C^∞ function $u_I^{(2)}$ that satisfies $\dfrac{\partial u_I^{(2)}}{\partial \overline{z}_{n-1}} = w'_I$ and is holomorphic with respect to z_n. Set $u^{(2)} := \sum_{\substack{I \not\ni n,n-1 \\ |I|=q-1}}' u_I^{(2)} d\overline{z}_I$; then $v - \overline{\partial} u^{(1)} - \overline{\partial} u^{(2)}$ contains neither $d\overline{z}_n$ nor $d\overline{z}_{n-1}$. In this way, if we keep producing new forms starting with v, then eventually the form reaches the $(0,q)$-form that does not contain any of $d\overline{z}_n, \cdots, d\overline{z}_1$, or the form 0 at which $v = \overline{\partial}\left(\sum_{j=1}^{n} u^{(j)}\right)$. □

COROLLARY 2.10. *$f \in C^\infty(\Omega)$ if both $f \in L^2_{\mathrm{loc}}(\Omega)$ and $\overline{\partial} f \in C^{0,1}(\Omega)$.*

PROOF. Since the equation $\overline{\partial} u = \overline{\partial} f$ for an unknown function u has locally a C^∞ solution, f is holomorphic modulo C^∞ functions. Hence, in particular, $f \in C^\infty(\Omega)$. □

Let us show one consequence of Serre's criterion.

PROPOSITION 2.11. *If $H^{0,q}(\Omega) = \{0\}$ for $1 \leqq q \leqq n-1$ and if $\partial\Omega$ is a real hypersurface of class C^1, then for every boundary point*

z_0 *of* Ω, *there exists an element* f *in* $A(\Omega)$ *that satisfies*

$$\overline{\lim_{z \to z_0}} |f(z)| = \infty.$$

PROOF. Let L be a complex line that intersects transversely with $\partial\Omega$ at z_0. Take a holomorphic function on $L \setminus \{z_0\}$ that has a pole at z_0, and extend it to Ω. Then this holomorphic extension f will satisfy the above condition. □

This proposition naturally poses the problem of relating the vanishing of $\overline{\partial}$ cohomology groups on Ω with such a geometric condition as the logarithmic convexity of complete Reinhardt domains.[3] For brevity, we call the condition

$$H^{0,q}(\Omega) = 0 \text{ for all } 1 \leqq q \leqq n-1$$

Serre's condition.

The $\overline{\partial}$ cohomology groups are related to the spectrum problem as follows:

THEOREM 2.12. *If* Ω *satisfies Serre's condition, then for every point* a *of* $\partial\Omega$, *there exist elements* $g_1, \cdots, g_n$ *in* $A(\Omega)$ *such that*

$$\sum_{j=1}^{n} (z_j - a_j)\, g_j(z) = 1.$$

COROLLARY 2.13. *An open set* Ω *that satisfies Serre's condition is closed in* $\mathrm{Spec}_m A(\Omega)$ *with respect to the weak topology.*

PROOF OF THEOREM 2.12. Let V be a module $C^\infty(\Omega)^{\oplus n}$ over $C^\infty(\Omega)$, and define a $C^\infty(\Omega)$ homomorphism $\alpha : V \to C^\infty(\Omega)$ by

$$\alpha(v_1, \cdots, v_n) = \sum_{j=1}^{n} z_j v_j.$$

Let $\{e_1, \cdots, e_n\}$ be the standard basis of V. In terms of α, define contractions $\alpha_k : \bigwedge^k V \to \bigwedge^{k-1} V$ pointwise by

$$\alpha_k(e_{i_1} \wedge \cdots \wedge e_{i_k}) = \sum_{j=1}^{k} (-1)^j \alpha(e_{i_j}) e_{i_1} \wedge \overset{\overset{i_j}{\vee}}{\cdots} \wedge e_{i_k}$$

and by linearity. Then the following exact sequence is obtained:

$$0 \longrightarrow \bigwedge^n V \xrightarrow{\alpha_n} \bigwedge^{n-1} V \xrightarrow{\alpha_{n-1}} \cdots \longrightarrow \bigwedge^2 V \xrightarrow{\alpha_2} V \xrightarrow{\alpha} C^\infty(\Omega) \longrightarrow 0.$$

[3]This will be described in detail starting in the next chapter.

If we extend the range of definition for the operator $\overline{\partial}$ to vector-valued differential forms, noting that α_k commutes with the operation of $\overline{\partial}$, then $\overline{\partial}$ cohomology groups $H^{p,q}(\operatorname{Ker}\alpha_k)$ with coefficients in $\operatorname{Ker}\alpha_k$ are defined by

$$\begin{aligned}&H^{p,q}(\operatorname{Ker}\alpha_k)\\&\quad := (C^{p,q}(\Omega)\otimes \operatorname{Ker}\alpha_k)\cap \operatorname{Ker}\overline{\partial}/\overline{\partial}\left(C^{p,q-1}(\Omega)\otimes \operatorname{Ker}\alpha_k\right).\end{aligned}$$ [4]

Then the same argument as for (2.2) induces that if $H^{0,k+1}(\operatorname{Ker}\alpha_{k+1}) = \{0\}$, then the homomorphism derived from α_{k+1},

$$H^{0,k}(\Omega)^{\oplus\binom{n}{k+1}} \longrightarrow H^{0,k}(\operatorname{Ker}\alpha_k),$$

becomes surjective.

Therefore, it follows that

$$\begin{aligned}\operatorname{Im}\,(\alpha \mid A(\Omega)^{\oplus n}) &= A(\Omega)\\ &\Longleftarrow H^{0,1}(\operatorname{Ker}\alpha)=\{0\}\\ &\Longleftarrow H^{0,1}(\Omega)=\{0\} \text{ and } H^{0,2}(\operatorname{Ker}\alpha_2)=\{0\}\\ &\quad\vdots\\ &\Longleftarrow H^{0,1}(\Omega)=H^{0,2}(\Omega)=\cdots=H^{0,n-1}(\Omega)=\{0\}\\ &\qquad \text{and } H^{0,n}(\operatorname{Ker}\alpha_n)=\{0\}.\end{aligned}$$

Since $\operatorname{Ker}\alpha_n = \{0\}$, the proof is complete. □

PROOF OF COROLLARY 2.13. If a sequence of points $\{z^{(\mu)}\}_{\mu=1}^{\infty} \subset \Omega$ is a convergent sequence with respect to the weak topology, then the $z^{(\mu)}$ are clearly a bounded sequence and must converge to some point inside Ω. Otherwise, as there is a subsequence $z^{(\mu_k)}$ that converges to some point of $\partial\Omega$, by choosing this point as a and taking $g_1, \cdots, g_n$ produced in Theorem 2.12, it would follow that

$$\lim_{k\to\infty}\sum_{j=1}^{n}|g_j(z^{(\mu_k)})| = \infty,$$

which contradicts the fact that the $z^{(\mu)}$ form a weakly convergent sequence. □

The above argument tells us that the smoothness assumption on $\partial\Omega$ in Proposition 2.11 is redundant. Let us emphasize this fact because of its importance:

[4] The tensor products on the right hand side are regarded as $C^\infty(\Omega)$-modules.

THEOREM 2.14. *If Serre's condition holds for Ω, then for every $z_0 \in \partial\Omega$ and every sequence $\{p_\mu\}$ of points in Ω that converges to z_0, there exists $f \in A(\Omega)$ such that*

$$\overline{\lim_{\mu\to\infty}} |f(p_\mu)| = \infty.$$

In general, an open set Ω of $\mathbb{C}^n$ is said to be a *domain of holomorphy*[5] if there is no connected open set U that possesses the following property:

(2.10) $U \not\subset \Omega$, and there is a nonempty open subset V of $\Omega \cap U$ such that for any element f in $A(\Omega)$, there is an element g in $A(U)$ that satisfies $f \mid V = g \mid V$.

Let us leave to the reader the verification that logarithmically convex Reinhardt domains, convex domains, and open sets in the complex plane are all domains of holomorphy.

By virtue of this terminology, it follows from Theorem 2.14 that *every open set satisfying Serre's condition is a domain of holomorphy.* The converse statement is also true, but the proof of it needs further preparation, and we postpone the details until the next chapter. Note that the following fact is derived from this converse statement, which we accept for the present.

PROPOSITION 2.15. *A necessary and sufficient condition for Ω to be a domain of holomorphy is that for any $z_0 \in \partial\Omega$ and any sequence $\{p_\mu\}$ of points in Ω that converges to z_0, there exists $f \in A(\Omega)$ such that*

$$\overline{\lim_{\mu\to\infty}} |f(p_\mu)| = \infty.$$

What we wanted to make clear in this chapter is that a region of definition Ω of holomorphic functions must be a domain of holomorphy in order to solve affirmatively the extension and division problems. Later on, we will indeed investigate whether these problems can be solved on domains of holomorphy. In deference to Kiyoshi Oka's methods, we will first characterize domains of holomorphy by the concept of pseudoconvexity, and then treat solutions of all the problems on domains of holomorphy as consequences of pseudoconvexity. Our lines are in imitation of what has been done repeatedly in mathematics, such as the reconstruction of Euclid's theory by means of Descartes' methods.

---

[5]Connectedness is not imposed on the definition of "domain" of holomorphy.

CHAPTER 3

Pseudoconvexity and Plurisubharmonic Functions

As seen in the preceding chapter, Ω must be a domain of holomorphy in order that the propositions corresponding to the Euclidean algorithm on the ring of integers and to Lagrange's interpolation on the ring of polynomials of one variable hold on the ring $A(\Omega)$ of holomorphic functions. On the other hand, these propositions are closely related to Serre's condition, the vanishing of $\overline{\partial}$ cohomology on Ω, and this condition provides a unified grasp of various phenomena on a domain of holomorphy. $\overline{\partial}$ cohomology is essentially under control of the pseudoconvexity of an open set. Pseudoconvexity is a concept similar to geometric convexity, but is much weaker as a condition.

This chapter begins with the definition of Hartogs pseudoconvexity and verifies that a domain of holomorphy is Hartogs pseudoconvex. Secondly, we show that some canonical function on a Hartogs pseudoconvex open set expressed by a distance function is plurisubharmonic. In consequence of this, it is derived that a domain of holomorphy is pseudoconvex. Hartogs and Oka's discovery of this relevancy gives a unique vitality to the theory of analytic functions of several variables.

In Chapter 4, in order to show that a pseudoconvex open set is a domain of holomorphy, some kind of differentiable plurisubharmonic functions will be needed. For this purpose, we detail the regularization of plurisubharmonic functions in the present chapter.

Finally, we mention the Levi pseudoconvexity and introduce basic facts and important examples of pseudoconvex open sets that have smooth boundaries. This also serves as an introduction to Chapter 6.

3.1. Pseudoconvexity of Domains of Holomorphy

A handhold is a domain

$$T_\varepsilon = \left\{(z_1, z_2) \in \Delta^2 \mid |z_1| < \varepsilon \text{ or } 1 - \varepsilon < |z_2| < 1\right\},$$

which is called a *Hartogs figure.*

DEFINITION 3.1. Ω is said to be *pseudoconvex in the sense of Hartogs*, or, for short, *Hartogs pseudoconvex*, if any holomorphic mapping from a Hartogs figure T_ε with ε arbitrary to Ω always extends to a holomorphic mapping from Δ^2 to Ω.

THEOREM 3.2. *$\mathbb{C}$ is Hartogs pseudoconvex.*

PROOF. It suffices to show that the restriction mapping $A(\Delta^2) \to A(T_\varepsilon)$ is surjective. Let $f \in A(T_\varepsilon)$. From Proposition 1.4, $f(z)$ expands into the power series

$$\sum_{k,l} c_{k,l} z_1^k z_2^l ,$$

which is convergent on $\Delta^2(0,(\varepsilon,1))$. After changing the order of the summation, let us observe the range on which the following equation holds:

$$f(z) = \sum_k \left(\sum_l c_{k,l} z_2^l \right) z_1^k .$$

By setting $c_k(z_2) := \sum\limits_l c_{k,l} z_2^l$ for $|z_2| < 1$, we regard the right hand side as a series with terms of holomorphic functions $c_k(z_2) z_1^k$. Since the left hand side is holomorphic on $\Delta \times \{z_2 \mid 1-\varepsilon < |z_2| < 1\}$, from the same argument as in (1.2) it follows that when $1-\varepsilon < |z_2| < 1$,

$$|c_k(z_2)| \leqq \sup_{|z_1|=r} |f(z_1,z_2)| r^{-k} \text{ for } 0 < r < 1. \tag{3.1}$$

Hence, from the maximum principle, when $|z_2| < r < 1$,

$$|c_k(z_2)| \leqq \sup_{|z_1|=|z_2|=r} |f(z_1,z_2)| r^{-k}. \tag{3.2}$$

Therefore, the series $\sum\limits_{k=1}^{\infty} c_k(z_2) z_1^k$ converges uniformly on compact sets in Δ^2, and Theorem 1.6 says that this series is a holomorphic extension of f to Δ^2 □

COROLLARY 3.3. *$\mathbb{C}^n$ is Hartogs pseudoconvex.*

THEOREM 3.4. *A domain of holomorphy is Hartogs pseudoconvex.*

PROOF. Let f be a holomorphic mapping from T_ε to Ω. From Corollary 3.3, f has a holomorphic extension $\widetilde{f} : \Delta^2 \to \mathbb{C}^n$. If it were true that $\widetilde{f}(\Delta^2) \not\subset \Omega$, then, letting $\widetilde{T}_\varepsilon$ be the maximum among connected open sets U that contains T_ε and satisfies $\widetilde{f}(U) \subset \Omega$, it would follow that $\partial\widetilde{T}_\varepsilon \cap \Delta^2 \neq \emptyset$. The maximum property of $\widetilde{T}_\varepsilon$ implies $\widetilde{f}(p) \in \partial\Omega$ for a given point $p \in \partial\widetilde{T}_\varepsilon \cap \Delta^2$. Hence, from Proposition 2.15, choosing a sequence of points p_μ in $\widetilde{T}_\varepsilon \cap \Delta^2$ that converges to p, there exists an element g in $A(\Omega)$ such that

$$\overline{\lim_{\mu\to\infty}} |g(f(p_\mu))| = \infty. \tag{3.3}$$

This contradicts the fact that $g \circ f$ has a holomorphic extension to Δ^2. □

COROLLARY 3.5. *The following are all Hartogs pseudoconvex open sets:*

1. *Logarithmically convex Reinhardt domains.*
2. *Convex domains.*
3. *Open sets in the complex plane.*

Next, observe a function $\delta_\Omega^v(z)$ on Ω defined by

$$\delta_\Omega^v(z) := \inf\left\{|\zeta| \;\middle|\; z + \frac{\zeta v}{\|v\|} \notin \Omega\right\}, \quad \text{where } v \in \mathbb{C}^n \setminus \{0\},$$

in order to relate Hartogs pseudoconvexity to a metric character. From the definition, $\delta_\Omega^v(z)$ is lower semicontinuous as a function from Ω to $(0,\infty]$. In addition to this, the following remarkable property of $\delta_\Omega^v(z)$ emerges if Ω is Hartogs pseudoconvex.

THEOREM 3.6 (F. Hartogs, K. Oka). *If Ω is Hartogs pseudoconvex, then for $z \in \Omega$, $v \in \mathbb{C}^n \setminus \{0\}, w \in \mathbb{C}^n$, and $r > 0$ that satisfy*

$$\left\{z + te^{i\theta}w \mid 0 \leqq t \leqq r \text{ and } 0 \leqq \theta < 2\pi\right\} \subset \Omega, \tag{3.4}$$

we have

$$\log \delta_\Omega^v(z) \geqq \frac{1}{2\pi}\int_0^{2\pi} \log \delta_\Omega^v(z + re^{i\theta}w)d\theta. \tag{3.5}$$

PROOF. Since $\log \delta_\Omega^v(z + re^{i\theta}w)$ is lower semicontinuous with respect to θ, this is the limit of an increasing sequence $\{\psi_R(\theta)\}_{R=1}^\infty$ of continuous functions. Set

$$u_R(te^{i\theta'}) := \frac{1}{2\pi}\int_0^{2\pi} \psi_R(\theta)\frac{r^2 - t^2}{r^2 - 2rt\cos(\theta - \theta') + t^2}d\theta\,;$$

then Fatou's theorem in the theory of Lebesgue integrals yields that

$$\lim_{R\to\infty} u_R(0) = \frac{1}{2\pi}\int_0^{2\pi} \log \delta^v_\Omega(z + re^{i\theta}w)d\theta. \tag{3.6}$$

Now take $h_R \in A(\Delta(0, r))$ with $u_R = \mathrm{Re}\ h_R$, and consider the mapping

$$\begin{array}{ccccc} \alpha_R : & \Delta^2 & \longrightarrow & \mathbb{C}^n \\ & \Psi & & \Psi \\ & (z_1, z_2) & \longmapsto & z + z_1 e^{h_R(z_2)} v + z_2 w. \end{array}$$

From the continuity of $\psi_R(\theta)$, we see that

$$\lim_{t\to r} u_R(te^{i\theta}) = \psi_R(\theta).$$

Therefore, given a positive number ε, we can choose an appropriate positive number δ such that

$$\{\alpha_R(z) \mid |z_1| < \delta \text{ or } 1 - \delta < |z_2| < 1 - \varepsilon\} \subset \Omega.$$

At this point, apply the Hartogs pseudoconvexity of Ω, then it follows that $\alpha_R(\Delta^2) \subset \Omega$.

Hence, in particular, we obtain $e^{u_R(0)} \leqq \delta^v_\Omega(z)$, or

$$u_R(0) \leqq \log \delta^v_\Omega(z). \tag{3.7}$$

From the combination of (3.6) and (3.7), the desired inequality (3.5) follows. □

For the reader's convenience, let us review the basics of subharmonic functions without proof. Let Ω be an open set in the complex plane for a while.

The following two conditions on an upper semicontinuous function ψ from Ω to $[-\infty, \infty)$ are equivalent:

(3.8) If $\overline{\Delta(z, r)} \subset \Omega$, then

$$\psi(z) \leqq \frac{1}{2\pi}\int_0^{2\pi} \psi(z + re^{i\theta})d\theta.$$

(3.9) If $\overline{\Delta(z, r)} \subset \Omega$, $h \in A(\overline{\Delta(z, r)})$, and

$$\mathrm{Re}\ h|\partial\Delta(z, r) \geqq \psi|\partial\Delta(z, r),$$

then

$$\mathrm{Re}\ h(z') \geqq \psi(z') \text{ for } z' \in \Delta(z, r).$$

When these conditions are met, ψ is called a *subharmonic function* on Ω. When $\overline{\Delta(z,r)} \subset \Omega$, given a subharmonic function ψ on Ω, define a function $M(\psi, t)$ by

$$M(\psi, t) := \frac{1}{2\pi} \int_0^{2\pi} \psi(z + te^{i\theta}) d\theta.$$

Then $M(\psi, t)$ is monotone increasing on t. This is obtained from (3.9) and the mean-value property of harmonic functions. We can take this condition on $M(\psi, t)$ as the definition of subharmonic function.

The next four properties follow immediately from the definition of subharmonic function:

(3.10) If φ and ψ are subharmonic, so are $\varphi + \psi$ and $\alpha\varphi$ ($\alpha \geqq 0$).

(3.11) For a family $\{\psi_\lambda\}_\lambda$ of subharmonic functions that are bounded from above uniformly on compact sets in Ω, set $\psi := \sup\limits_{\lambda} \psi_\lambda$ and $\psi^*(z) := \lim\limits_{\varepsilon \to 0} \sup\limits_{z' \in \mathbb{B}(z,\varepsilon)} \psi(z')$. Then ψ^* is subharmonic on Ω.

(3.12) A real-valued function ψ of class C^2 on Ω is subharmonic if and only if the following differential inequality holds everywhere:

$$\frac{\partial^2 \psi}{\partial z \partial \overline{z}} \geqq 0.$$

(3.13) If a sequence $\{\psi_j\}_{j=1}^\infty$ of subharmonic functions on Ω is monotone decreasing, then $\lim\limits_{j \to \infty} \psi_j$ is subharmonic.

A fact that is immediate from (3.12) and often applied is that if ψ is subharmonic and of class C^2, then $\lambda(\psi)$ is subharmonic for any increasing convex function λ of class C^2 on $\mathbb{R}$. Combining this with (3.13), for instance, it is easily seen that $\log \sum\limits_{j=1}^m |f_j|^2$ and $\left(\sum\limits_{j=1}^m |f_j|^2 \right)^\alpha$ ($\alpha \geqq 0$) are subharmonic on Ω for $f_1, \cdots, f_m \in A(\Omega)$.

Let us return to the topics on $\mathbb{C}^n$. The formula (3.5) indicates the subharmonicity of $-\log \delta_\Omega^v(z + \zeta w)$ with respect to ζ. Namely, $-\log \delta_\Omega^v$ is subharmonic on $L \cap \Omega$ for any complex line L.

DEFINITION 3.7. An upper semicontinuous function

$$\psi \colon \Omega \to [-\infty, \infty)$$

is said to be *plurisubharmonic* if $\psi(z+\zeta w)$ is subharmonic as a function of ζ for a given $(z,w) \in \Omega \times \mathbb{C}^n$.

Denote by $\mathrm{PSH}(\Omega)$ the set of all plurisubharmonic functions on Ω. For simplicity, Ω may be omitted from this notation.

The basics of plurisubharmonicity are totally the same as in the case of subharmonic functions, but we list them for convenience:

(3.14) If φ and ψ are in $\mathrm{PSH}(\Omega)$, so are $\varphi+\psi$ and $\alpha\varphi$ $(\alpha \geqq 0)$.

(3.15) If $\psi \in \mathrm{PSH}(\Omega)$, then, given $z \in \Omega$,

$$M(\psi,t) := \frac{1}{\mathrm{Vol}\,\partial\mathbb{B}(z,t)} \int_{\partial\mathbb{B}(z,t)} \psi\, dS$$

is monotone increasing on t with $\overline{\mathbb{B}(z,t)} \subset \Omega$.

(3.16) If $\{\psi_\lambda\}_{\lambda\in\Lambda} \subset \mathrm{PSH}(\Omega)$ and if ψ_λ are bounded above uniformly on compact sets in Ω, then

$$\psi^*(z) = \lim_{\varepsilon\to 0} \sup_{z'\in\mathbb{B}(z,\varepsilon)} \sup_\lambda \psi_\lambda(z') \in \mathrm{PSH}(\Omega).$$

(3.17) Given a real-valued function ψ of class C^2 on Ω, $\psi \in \mathrm{PSH}(\Omega)$ if and only if the $n\times n$ matrix

$$\left(\frac{\partial^2\psi}{\partial z_j \partial \bar{z}_k}\right)$$

is semipositive definite everywhere on Ω.

(3.18) If a sequence of functions $\{\psi_j\}_{j=1}^\infty \subset \mathrm{PSH}(\Omega)$ is monotone decreasing, then $\lim_{j\to\infty} \psi_j \in \mathrm{PSH}(\Omega)$.

From Theorem 3.6 and (3.16), it is seen that

$$\Omega \text{ is Hartogs pseudoconvex} \implies -\log\delta_\Omega \in \mathrm{PSH}(\Omega),$$

where $\delta_\Omega(z) := \inf_{w\in\partial\Omega} |z-w|$. Unlike δ_Ω^v, δ_Ω is a continuous function that has some finite determinate values on Ω as long as $\Omega \neq \mathbb{C}^n$.

DEFINITION 3.8. Ω is said to be *pseudoconvex* if there exists a continuous plurisubharmonic function $\psi\colon \Omega \to \mathbb{R}$ such that the set

$$\Omega_{\psi,c} := \{z \in \Omega \mid \psi(z) < c\}$$

is relatively compact inside Ω for each $c \in \mathbb{R}$.

It is clear from the definition and either (3.14) or (3.16) that if Ω_1 and Ω_2 are pseudoconvex, then so is $\Omega_1 \cap \Omega_2$. Pseudoconvexity and Hartogs pseudoconvexity are equivalent, but before giving the proof, we will describe the regularization of plurisubharmonic functions in the next section.

REMARK. From the proof of Theorem 3.6, it turns out that if Ω is not pseudoconvex, then there exists a biholomorphic mapping ι from Δ^n onto an open set $U \subset \mathbb{C}^n$ such that

$$\iota(T_\varepsilon \times \Delta^{n-2}) \subset \Omega \quad \text{and} \quad U \not\subset \Omega.$$

This means, by Theorem 3.2, that all elements in $A(\Omega)$ extend to U as holomorphic functions. But the value of the analytic continuation of a holomorphic function does not necessarily coincide with the value of the original function at a point in $\Omega \cap U$ except $\iota(T_\varepsilon \times \Delta^{n-2})$. Therefore, in order to discuss, in general, the global theory of holomorphic functions, it would be insufficient to restrict the domains of functions to open sets in $\mathbb{C}^n$. However, as long as we stay only within the theory explained in this book, the replacement of Ω with a more general space such as a complex manifold will not affect the framework of the theory (although this replacement would put things in much wider perspective). Hence, in order to give priority to brevity, we will be restricting ourselves to open subsets of $\mathbb{C}^n$ from now on, too.

3.2. Regularization of Plurisubharmonic Functions

The existence of differentiable plurisubharmonic functions is important, as we will need to differentiate formulae that contain plurisubharmonic functions later on to solve the $\overline{\partial}$ equation on pseudoconvex open sets. As preparation for this, we describe below the regularization of plurisubharmonic functions.

Let Ω_ε and μ_ε be as defined after (1.9) in § 1. 2.

Let ψ be a plurisubharmonic function that is locally integrable on Ω, and set

$$\psi_\varepsilon(z) := \int_{\mathbb{C}^n} \psi(z+\zeta)\mu_\varepsilon(\zeta)dV_\zeta \text{ for } z \in \Omega_\varepsilon\,.$$

Clearly, $\psi_\varepsilon \in C^\infty(\Omega_\varepsilon)$. Furthermore, ψ_ε has the following property:

PROPOSITION 3.9. $\psi_\varepsilon \in \mathrm{PSH}(\Omega_\varepsilon)$, *and* $\psi_\varepsilon \searrow \psi$ *as* $\varepsilon \searrow 0$.

PROOF. The monotonicity of the family of functions ψ_ε follows from that of $M(\psi, t)$. The plurisubharmonicity of ψ_ε is shown as follows:

$$\begin{aligned}
&\frac{1}{2\pi}\int_0^{2\pi}\psi_\varepsilon(z+re^{i\theta}w)d\theta \\
&=\frac{1}{2\pi}\int_{\mathbb{C}^n}\mu_1(\zeta)dV_\zeta\int_0^{2\pi}\psi(z+re^{i\theta}w-\varepsilon\zeta)d\theta \\
&\geqq\int_{\mathbb{C}^n}\mu_1(\zeta)\psi(z-\varepsilon\zeta)dV_\zeta \\
&=\psi_\varepsilon(z),
\end{aligned}$$

where r is chosen to be sufficiently small. □

COROLLARY 3.10. *Let $\psi \in \mathrm{PSH}(\Omega)$. If λ is an increasing convex function (in the broad sense) defined on an interval that contains the range of ψ, then $\lambda(\psi) \in \mathrm{PSH}(\Omega)$.*

PROOF. There is a family of increasing convex functions λ_μ of class C^∞ with $\lambda_\mu \searrow \lambda$. For this family, (3.17) implies $\lambda_\mu(\psi_\varepsilon) \in \mathrm{PSH}(\Omega_\varepsilon)$ by direct differentiation, while since $\lambda_\mu(\psi_\varepsilon) \searrow \lambda(\psi)$, (3.18) yields $\lambda(\psi) \in \mathrm{PSH}(\Omega)$. □

COROLLARY 3.11. *Let $\Omega_1 \subset \mathbb{C}^m$ and $\Omega_2 \subset \mathbb{C}^n$ be open sets, and let $F\colon \Omega_1 \to \Omega_2$ be a holomorphic mapping. Then, for any $\psi \in \mathrm{PSH}(\Omega_2)$, one has that $\psi \circ F \in \mathrm{PSH}(\Omega_1)$.*

PROOF. If $\psi \in C^2(\Omega_2) \cap \mathrm{PSH}$, then $\psi \circ F \in \mathrm{PSH}(\Omega_1)$ follows from taking its derivatives. For the general case, use an approximate family as in Corollary 3.10. □

THEOREM 3.12. *Hartogs pseudoconvexity and pseudoconvexity are equivalent to each other.*

PROOF. Hartogs pseudoconvexity $\Longrightarrow$ Pseudoconvexity: As ψ, take $|z|^2$ for the case $\Omega = \mathbb{C}^n$, and $|z|^2 - \log \delta_\Omega$ otherwise.

Psuedoconvexity $\Longrightarrow$ Hartogs pseudoconvexity: From the property (3.15) of plurisubharmonic functions and Corollary 3.11, it suffices to repeat an argument similar to the proof of Theorem 3.4. □

COROLLARY 3.13. *For an increasing sequence $\{\Omega_k\}_{k=1}^\infty$ of pseudoconvex open sets, $\bigcup_{k=1}^\infty \Omega_k$ is pseudoconvex.*

If plurisubharmonic functions are continuous, a stronger approximation theorem holds. To describe this, let us introduce the concept of strictly plurisubharmonic function.

In general, given a locally integrable function φ on Ω and $z_0 \in \Omega$, let $L[\varphi](z_0)$ ($\in [-\infty, \infty]$) be the supremum of real numbers ε such that $\varphi(z) - \varepsilon|z|^2$ is plurisubharmonic on some neighborhood of z_0. Clearly,

$$L[\varphi + \psi] \geqq L[\varphi] + L[\psi]$$

and

$$L[\max(\varphi, \psi)] \geqq \min(L[\varphi], L[\psi]).$$

It is also obvious that $L[\varphi]$ is lower semicontinuous.

If $L[\psi](z_0) > 0$, ψ is said to be *strictly plurisubharmonic* at z_0, and if $L[\psi] > 0$ on Ω, ψ is called a *strictly plurisubharmonic function* on Ω.[1]

Denote by $\mathrm{PSH}^*(\Omega)$ the set of all strictly plurisubharmonic functions on Ω.

THEOREM 3.14 (Richberg's theorem). *If $\psi \in \mathrm{PSH}^*(\Omega) \cap C^0(\Omega)$, then for any positive-valued continuous function ε on Ω, there exists a function $\varphi \in \mathrm{PSH}^*(\Omega) \cap C^\infty(\Omega)$ that satisfies the inequalities*

$$\psi \leqq \varphi < \psi + \varepsilon.$$

PROOF. Take countably many n-dimensional open balls $\mathbb{B}(p_j, R_j)$ ($j = 1, 2, \cdots$) that satisfy all the following three conditions:

(3.19) $\overline{\mathbb{B}(p_j, R_j)} \subset \Omega$.

(3.20) $\bigcup_{j=1}^{\infty} \mathbb{B}\left(p_j, \frac{R_j}{2}\right) = \Omega$.

(3.21) For an arbitrary compact set K in Ω, there are only finitely many j's such that $\mathbb{B}(p_j, R_j) \cap K \neq \emptyset$.

Also, fix an increasing convex function λ (in the broad sense) of class C^∞ on $\mathbb{R}$ such that $\operatorname{supp} \lambda \subset \left(\frac{1}{3}, \infty\right)$ and $\lambda(1) = 1$.

For ψ and ε, let us construct inductively an element $\psi_{(k)}$ in $\mathrm{PSH}^*(\Omega)$ that is of class C^∞ on a neighborhood of $\overline{\bigcup_{j=1}^{k} \mathbb{B}\left(p_j, \frac{R_j}{2}\right)}$ and satisfies $\psi \leqq \psi_{(k)} < \psi + \varepsilon$ on Ω.

For the case $k = 1$, since $\psi \in \mathrm{PSH}^*(\Omega)$, there is a positive number η_1 such that

$$L\left[\psi - \eta_1 \lambda\left(\frac{|z - p_1|^2}{R_1^2}\right)\right] > \frac{2}{3} L[\psi] \tag{3.22}$$

[1] In the literature, the condition "of class C^2" is also frequently included.

on $\mathbb{B}(p_1, R_1)$. Hence, given a positive number δ, set

$$\psi^\delta := \psi_\delta - \eta_1 \lambda\left(\frac{|z-p_1|^2}{R_1^2}\right);$$

then, for a sufficiently small δ,

$$L[\psi^\delta] > \frac{2}{3} L[\psi] \text{ and } \psi^\delta < \psi + \varepsilon \tag{3.23}$$

on $\mathbb{B}(p_1, R_1)$. In addition, we have

$$\psi^\delta < \psi \tag{3.24}$$

on some neighborhood of $\partial\mathbb{B}(p_1, R_1)$. Fix such a δ, and set

$$\psi_{(1)}(z) := \begin{cases} \max\{\psi(z), \psi^\delta(z)\} & \text{for } z \in \mathbb{B}(p_1, R_1), \\ \psi(z) & \text{for } z \in \Omega \setminus \mathbb{B}(p_1, R_1). \end{cases} \tag{3.25}$$

Then $\psi \leqq \psi_{(1)} < \psi + \varepsilon$ and $\psi_{(1)}\Big|\mathbb{B}\left(p_1, \frac{R_1}{2}\right) \in C^\infty(\Omega)$. From the construction, it is also obvious that $L[\psi_{(1)}] > \frac{2}{3} L[\psi]$ on $\mathbb{B}(p_1, R_1)$.

The method of producing $\psi_{(k+1)}$ from $\psi_{(k)}$ is described as follows:

First, by a method similar to the above, transform $\psi_{(k)}$ to $\psi^\delta_{(k)}$ on $\mathbb{B}(p_{k+1}, R_{k+1})$. But in this process, replace the condition (3.22) by

$$L\left[\psi_{(k)} - \eta_{k+1}\lambda\left(\frac{|z-p_{k+1}|^2}{R_{k+1}^2}\right)\right] > \left(1 - \sum_{j=1}^{k+1} \frac{1}{3^j}\right) L[\psi], \tag{3.26}$$

so that $\psi^\delta_{(k)}$ possesses the properties corresponding to those of ψ^δ on $\mathbb{B}(p_{k+1}, R_{k+1})$.

Let χ_k be a nonnegative real-valued function of class C^∞ on Ω with a compact support such that $\chi_k\Big|\bigcup_{j=1}^{k} \mathbb{B}\left(p_j, \frac{R_j}{2}\right) \equiv 1$ and $\psi_{(k)}$ is of class C^∞ on $\operatorname{supp}\chi_k$. Set

$$\psi_{(k+1)}(z) := \begin{cases} \max\{\psi_{(k)}(z), (1-\chi_k(z))\psi^\delta_{(k)}(z) + \chi_k(z)\psi_{(k)}(z)\} \\ \qquad\qquad \text{for } z \in \mathbb{B}(p_{k+1}, R_{k+1}), \\ \psi_{(k)}(z) \qquad \text{for } z \in \Omega \setminus \mathbb{B}(p_{k+1}, R_{k+1}). \end{cases} \tag{3.27}$$

Clearly, $\psi_{(k+1)} \geqq \psi_{(k)} \geqq \psi$. By taking a sufficiently small δ, we get

(3.28)

$$L[(1-\chi_k)\psi^{\delta}_{(k)} + \chi_k \psi_{(k)}] > \left(1 - \sum_{j=1}^{k+1} \frac{1}{3^j}\right) L[\psi] \text{ and } \psi_{(k+1)} < \psi + \varepsilon$$

on $\mathbb{B}(p_{k+1}, R_{k+1})$, and furthermore

(3.29) $$\psi^{\delta}_{(k)} < \psi_{(k)}$$

on some neighborhood of $\partial\mathbb{B}(p_{k+1}, R_{k+1})$. In this case, $\psi_{(k+1)}$ coincides with $\psi_{(k)}$ on some neighborhood of $\Omega \setminus \mathbb{B}(p_{k+1}, R_{k+1})$. Also, since $\psi^{\delta}_{(k)} \geqq \psi_{(k)}$ on $\mathbb{B}\left(p_{k+1}, \dfrac{R_{k+1}}{2}\right)$, it follows that $\psi_{(k+1)}$ coincides with $(1-\chi_k)\psi^{\delta}_{(k)} + \chi_k\psi_{(k)}$ and is of class C^∞ on this open ball. Hence, all the requirements are met.

From the construction, we have $\psi_{(k+1)} = \psi_{(k)}$ on $\bigcup_{j=1}^{k} \mathbb{B}\left(p_j, \dfrac{R_j}{2}\right)$. Therefore, $\varphi := \lim_{k\to\infty} \psi_{(k)}$ exists, satisfying $\psi \leqq \varphi < \psi + \varepsilon$ and $\varphi \in \mathrm{PSH}^* \cap C^\infty(\Omega)$. □

In general, a real-valued continuous function φ on a topological space X is said to be an *exhaustion function* on X if the subset $\{x \mid \varphi(x) < c\}$ of X is relatively compact for every real number c that is less than the supremum of the values of φ.

By virtue of Richberg's theorem, the definition of pseudoconvexity can be strengthened as follows:

THEOREM 3.15. *A pseudoconvex open set has an unbounded strictly plurisubharmonic exhaustion function of class C^∞.*

In recent years a remarkable result on the approximation of plurisubharmonic functions has been obtained. For the present, let us introduce the statement of this result while putting off its proof.

Let ψ be a plurisubharmonic function on Ω. The *Lelong number* $\nu(\psi, x_0)$ of ψ at a point x_0 of Ω is defined by

$$\nu(\psi, x_0) := \liminf_{z\to x_0} \frac{\psi(z)}{\log|z-x_0|} \quad \left(= \lim_{r\searrow 0} \frac{\sup_{\mathbb{B}(x_0,r)} \psi}{\log r}\right).$$

Recall that $\log|f| \in \mathrm{PSH}(\Omega)$ for $f \in A(\Omega)$. It is easy to see that

$$\nu(\log|f|, x_0) = \sup\{k \in \mathbb{Z}_+ \mid \langle\alpha\rangle < k \text{ implies } f^{(\alpha)}(x_0) = 0\}.$$

For a locally integrable plurisubharmonic function ψ, define a Hilbert space

$$A^2_{2m\psi}(\Omega) := \left\{ f \in A(\Omega) \;\Big|\; \int_\Omega |f|^2 e^{-2m\psi} dV < \infty \right\}.$$

Take an orthonormal basis $\{\sigma_l\}_{l=1}^\infty$ of $A^2_{2m\psi}(\Omega)$, and set

$$\psi_m := \frac{1}{2m} \log \sum_{l=1}^{\infty} |\sigma_l|^2.$$

THEOREM 3.16 (J.-P. Demailly, 1992). *There exist constants C_1 and C_2, independent of m, which satisfy the following conditions:*

a. $\psi(z) - \dfrac{C_1}{m} \leqq \psi_m(z) \leqq \sup\limits_{|\zeta - z| < r} \psi(\zeta) + \dfrac{1}{m} \log \dfrac{C_2}{r^n}$, *where $z \in \Omega$ and $r < \delta_\Omega(z)$.*

b. $\nu(\psi, z) - \dfrac{n}{m} \leqq \nu(\psi_m, z) \leqq \nu(\psi, z)$, *where $z \in \Omega$.*

As an application of Demailly's theorem (whose proof will be given in § 5. 4 (b)), we immediately obtain a deep result on the Lelong number.

COROLLARY 3.17 (Siu's theorem). *Let Ω and ψ be as defined above. Then, given a positive number c, the set*

$$E_c(\psi) := \{z \in \Omega \mid \nu(\psi, z) \geqq c\}$$

is an analytic subset of Ω.

PROOF. From Theorem 3.16 (b),

$$E_c(\psi) = \bigcap_{m \geqq 1} E_{c-n/m}(\psi_m).$$

However, since

$$E_{c-n/m}(\psi_m) \\ = \{z \mid \langle\alpha\rangle < mc - n \text{ implies } \sigma_l^{(\alpha)}(z) = 0 \text{ for } l = 1, 2, \cdots\},$$

it follows that the $E_{c-n/m}(\psi_m)$ are analytic subsets, and so is the intersection $E_c(\psi)$ of these sets. □

REMARK. $E_c(\psi)$ is clearly monotone decreasing with respect to c, but not much more than this is known. From a property of analytic subsets, it has been known that $E_c(\psi)$ is left continuous on c and has only countably many discontinuous points. At a discontinuous point c', there occurs the phenomenon that a family of analytic subsets of

$E_c(\psi)$ as $c \searrow c'$ is absorbed by a higher-dimensional analytic subset in $E_{c'}(\psi)$.

REMARK. The first conjecture on the approximation of plurisubharmonic functions was made by Bochner and Martin [**4**, p.145] in relation to the Levi problem in the following form:

DEFINITION 3.18. A *Hartogs function* on Ω is by definition an element of the smallest among the families $\mathcal{F}(\Omega)$ of functions on Ω with values in $\mathbb{R} \cup \{-\infty\}$ that satisfy the following conditions:

1. $f \in A(\Omega)$ implies $\log|f| \in \mathcal{F}(\Omega)$.
2. $f_1, f_2 \in \mathcal{F}(\Omega)$ and $c \in [0, \infty)$ imply $f_1 + f_2, cf_1 \in \mathcal{F}(\Omega)$.
3. If $\{f_\lambda\}_{\lambda \in \Lambda} \subset \mathcal{F}(\Omega)$, and if the f_λ's are uniformly bounded from above on compact subsets in Ω, then $\sup\limits_{\lambda} f_\lambda \in \mathcal{F}(\Omega)$.
4. Both $\{f_j\}_{j=1}^{\infty} \subset \mathcal{F}(\Omega)$ and $f_j \geqq f_{j+1}$ imply $\lim\limits_{j \to \infty} f_j \in \mathcal{F}(\Omega)$.
5. $f \in \mathcal{F}(\Omega)$ implies $\overline{\lim\limits_{z' \to z}} f(z') \in \mathcal{F}(\Omega)$.
6. Letting $\mathcal{H}(\Omega)$ be the smallest of the $\mathcal{F}(\Omega)$'s that satisfy (1) – (5), if f satisfies $f \mid \Omega^* \in \mathcal{H}(\Omega^*)$ for any relatively compact open set Ω^* of Ω, then $f \in \mathcal{F}(\Omega)$.

Bochner–Martin conjecture:
Plurisubharmonic functions should be Hartogs functions.

If Ω is pseudoconvex, we see that the B–M conjecture is correct by using the solution of the Levi problem (Bremermann [**6**]). On the other hand, clearly the B–M conjecture would affirmatively solve the Levi problem, but there is a domain that gives a counterexample against the B–M conjecture ([ibid.]).

Roughly speaking, Bremermann's result corresponds to the polyhedral approximation of figures, while Demailly's theorem corresponds to the approximation by surfaces.

3.3. Levi Pseudoconvexity

Open sets that have boundaries of class C^2 are mainly described. A boundary point p of Ω is said to be *of class* C^k if there exist a neighborhood U of p in $\mathbb{C}^n$ and a real-valued function r_U of class C^k on U that satisfy the following two conditions:

$$U \cap \Omega = \{z \mid r_U(z) < 0\}. \tag{3.30}$$

$$\{z \in U \mid dr_U(z) = 0\} = \emptyset. \tag{3.31}$$

In this case, r_U is called a *defining function* of Ω on U or simply around p. When every point of $\partial\Omega$ is of class C^k, Ω is said to have *boundary of class* C^k, and we write $\partial\Omega \in C^k$. By means of a partition of unity, we can construct a real-valued function r of class C^k defined on a neighborhood of $\overline{\Omega}$ that satisfies

$$\Omega = \{z \mid r(z) < 0\}, \tag{3.32}$$

$$\{z \in \partial\Omega \mid dr(z) = 0\} = \emptyset\,. \tag{3.33}$$

In general, we call such a function r a *defining function* of Ω.

Let p be a C^2 boundary point of Ω. Since the tangent space to $\partial\Omega$ at p is of real dimension $2n-1$, this space contains a complex hyperplane. This is called a *complex tangent space* of $\partial\Omega$ at p, and is denoted by T'_p. Given a defining function r of Ω around p, the quadratic form

$$L_r(\xi, \xi) := \sum_{j,k} \frac{\partial^2 r}{\partial z_j \partial \overline{z}_k}(p)\xi_j\overline{\xi}_k \tag{3.34}$$

on the vector space $\{\xi \in \mathbb{C}^n \mid \xi + p \in T'_p\}$ ($= \operatorname{Ker} \partial r(p)$) is said to be the *Levi form* of r at p.

DEFINITION 3.19. A C^2 boundary point p of Ω is said to be *strongly pseudoconvex* if either $n = 1$ or the Levi form of r at p is positive definite.

Ω is called a *strongly pseudoconvex open set* if Ω is bounded and its boundary points are all strongly pseudoconvex.

The positive definiteness of the Levi form does not depend on the choice of a defining function r, since another defining function is written as a multiple ur by a positive-valued function u of class C^1, and, given $p \in \partial\Omega$, we have

$$\partial(ur)(p) = (u\partial r)(p), \tag{3.35}$$

and

$$\partial\overline{\partial}(ur)(p) = (u\partial\overline{\partial}r + \partial u \wedge \overline{\partial}r + \partial r \wedge \overline{\partial}u)(p). \tag{3.36}$$

In particular, take $u = e^{Br}$; then the right hand side of (3.36) becomes $(\partial\overline{\partial}r + 2B\partial r \wedge \overline{\partial}r)(p)$. Hence, if the Levi form of r is positive definite at p, then ur is strictly plurisubharmonic near p for a sufficiently large B.

From this the next proposition follows.

PROPOSITION 3.20. *A strongly pseudoconvex open set is pseudoconvex.*

PROOF. For a strongly pseudoconvex open set Ω, from the above argument, there is a defining function r of Ω that is strictly plurisubharmonic on a neighborhood of $\partial\Omega$. Since Ω is bounded, there is a sufficiently large positive number C such that $-\log(-r) + C|z|^2$ is exhaustive and strictly plurisubharmonic on Ω. □

From (3.36) it follows that the signature of the Levi form does not depend on the choice of a defining function.

Ω is said to be *Levi pseudoconvex* at p if the Levi form of a defining function is semipositive definite at p. An open set whose boundary points are all Levi pseudoconvex is called a *Levi pseudoconvex open set*.

THEOREM 3.21. *A Levi pseudoconvex open set is pseudoconvex.*

PROOF. For $p \in \partial\Omega$, set

$$\Omega_{p,\varepsilon} := \{z \mid r(z) + \varepsilon|z-p|^2 < 0\}, \text{ where } \varepsilon > 0.$$

From the assumption, there are a neighborhood U of p and a positive number ε_0 that satisfy the following two conditions:

1. Given $z \in U$ and $\varepsilon < \varepsilon_0$, there is a point w in $\partial\Omega_{p,\varepsilon} \cap U$ such that $\delta_{\Omega_{p,\varepsilon}}(z) = |z-w|$.
2. Given $\varepsilon < \varepsilon_0$ and $q \in \partial\Omega_{p,\varepsilon} \cap U$, the Levi form of $r + \varepsilon|z-p|^2$ is positive definite at q.

In this case, since every point of $\partial\Omega_{p,\varepsilon} \cap U$ is a strongly pseudoconvex boundary point of $\Omega_{p,\varepsilon}$, it follows that $-\log\delta_{\Omega_{p,\varepsilon}}$ is plurisubharmonic on $\Omega_{p,\varepsilon} \cap U$. Moreover, since $-\log\delta_{\Omega_{p,\varepsilon}} \searrow -\log\delta_\Omega$ as $\varepsilon \searrow 0$, (3.18) implies that $-\log\delta_\Omega$ is plurisubharmonic on $\Omega \cap U$. As p was chosen arbitrarily, there is some neighborhood W of $\partial\Omega$ such that $-\log\delta_\Omega \in \mathrm{PSH}(\Omega \cap W)$. Finally, compose $-\log\delta_\Omega$ with an appropriate increasing convex function; then we obtain a plurisubharmonic exhaustion function on Ω. (See Corollary 3.11.) □

That the Levi pseudoconvexity is a fundamental concept is also understood from the following:

PROPOSITION 3.22. *A pseudoconvex open set that has a C^2 boundary is Levi pseudoconvex.*

PROOF. Let ψ be a plurisubharmonic exhaustion function on Ω. If there were a boundary point p at which Ω is not Levi pseudoconvex, then there would be some $v \in T'_p$ such that

$$\sum_{j,k} \frac{\partial^2 r}{\partial z_j \partial \overline{z}_k}(p)(v-p)_j \overline{(v-p)_k} < 0. \tag{3.37}$$

Let ν_p be the unit inward normal vector to $\partial\Omega$ at p, and consider a family of holomorphic mappings:

$$\begin{array}{rcccl} \pi_t : & \mathbb{C} & \longrightarrow & \mathbb{C}^n & \\ & \Psi & & \Psi & \\ & \zeta & \longmapsto & p + t\nu_p + \zeta \cdot (v-p) & \quad \text{for } t \geqq 0. \end{array}$$

Then from (3.37) there is a positive number ε such that

$$\begin{cases} \pi_0(\overline{\Delta(0,\varepsilon)} \setminus \{0\}) \subset \Omega, \\ \pi_t(\overline{\Delta(0,\varepsilon)}) \subset \Omega \text{ for } 0 < t \leqq \varepsilon. \end{cases} \tag{3.38}$$

Note that $\psi \circ \pi_t$ is subharmonic on its domain, since ψ is plurisubharmonic. From this, it follows that

$$\begin{aligned} \sup_{\Omega} \psi &= \sup\{\psi \circ \pi_t(\zeta) \mid \zeta \in \Delta(0,\varepsilon) \text{ for } 0 < t \leqq \varepsilon\} \\ &\leqq \sup\{\psi \circ \pi_t(\zeta) \mid \zeta \in \partial\Delta(0,\varepsilon) \text{ for } 0 \leqq t \leqq \varepsilon\} \\ &< \sup_{\Omega} \psi, \end{aligned}$$

which is a contradiction. Therefore, Ω must be Levi pseudoconvex. □

It is evident that Levi pseudoconvexity does not necessarily imply strong pseudoconvexity, but we call attention to the following two facts:

PROPOSITION 3.23. *If $\partial\Omega \in C^2$ and Ω is bounded, then Ω has a strongly pseudoconvex boundary point.*

PROOF. It suffices to set $R := \sup_{z \in \Omega} |z|$ and choose a point p of $\partial\Omega$ such that $|p| = R$. □

PROPOSITION 3.24. *If a pseudoconvex domain Ω is bounded, and $\partial\Omega \in C^\omega$, then the set of all the strongly pseudoconvex boundary points of Ω is a dense open subset of $\partial\Omega$.*

PROOF. If $n = 1$, from the definition Ω is strongly pseudoconvex. In the case $n \geqq 2$, $\partial\Omega$ must be connected. Otherwise, there would be a most inside component of $\partial\Omega$, and from Proposition 3.23 its inside would have a strongly pseudoconvex boundary point. This means that the same point cannot be a Levi pseudoconvex boundary point of Ω, hence a contradiction. Now that $\partial\Omega$ is connected, has a strongly pseudoconvex boundary point, and is real analytic, the desired consequence follows from the theorem of identity. □

The two examples stated below are well-known, and explain much about general Levi pseudoconvex domains.

EXAMPLE 3.25 (Kohn–Nirenberg). Set

$$\Omega_{\mathrm{KN}} := \left\{ z \in \mathbb{C}^2 \;\middle|\; r := \mathrm{Re}\ z_2 + |z_1|^8 + \frac{15}{7}|z_1|^2\,\mathrm{Re}\ z_1^6 < 0 \right\}.$$

Since $r = \mathrm{Re}\ z_2 + z_1^4\overline{z}_1^4 + \frac{15}{14}(z_1^7\overline{z}_1 + \overline{z}_1^7 z_1)$, it follows that

$$r_{z_2\overline{z}_2} = r_{z_1\overline{z}_2} = r_{z_2\overline{z}_1} = 0,$$

and

$$\begin{aligned} r_{z_1\overline{z}_1} &= 16|z_1|^6 + 15\mathrm{Re}\ z_1^6 \\ &\geqq 16|z_1|^6 - 15|z_1|^6 = |z_1|^6 \geqq 0\,. \end{aligned}$$

Hence, $r \in \mathrm{PSH}(\mathbb{C}^2)$, and Ω_{KN} is pseudoconvex. Moreover, $\partial\Omega_{\mathrm{KM}}$ is of class C^ω on some neighborhood of 0, since $0 \in \partial\Omega_{\mathrm{KM}}$ and $\partial r(0) = dz_2(0)$. Therefore, from Proposition 3.22, Ω_{KM} is Levi pseudoconvex. Notice that Ω_{KM} shows, at first glance, a singular property at 0 as follows:

(3.39) If $f \in A(\mathbb{B}(0,\varepsilon))$ with $\varepsilon > 0$ and $f(0) = 0$, then

$$V(f) \cap \Omega_{\mathrm{KN}} \neq \emptyset.$$

For the proof of this fact, we refer the reader to either [**20**] or [**31**].

The Kohn–Nirenberg example presents a striking contrast to the following self-evident fact:

PROPOSITION 3.26. *If p is a strongly pseudoconvex boundary point of Ω, then there exist a neighborhood U of p and a biholomorphic mapping $\pi\colon U \to \Delta^n$ such that $\pi(p) = 0$ and*

$$\pi(U \cap \Omega) \subset \left\{ z \;\middle|\; \mathrm{Re}\ z_n + \sum_{j=1}^{n-1} |z_j|^2 < 0 \right\}.$$

In particular, putting $f := z_n \circ \pi$, *we obtain* $f \in A(U)$, $f(p) = 0$, *and* $V(f) \cap \Omega = \emptyset$.

This means that 'strong pseudoconvexity $\equiv$ convexity in the narrow sense' modulo biholomorphic equivalence, and Ω_{KN} provides an example showing that 'strong' cannot be replaced by 'Levi,' and 'narrow' cannot be replaced by 'broad.'

EXAMPLE 3.27 (Diederich–Fornaess). Let $\lambda\colon \mathbb{R} \to \mathbb{R}$ be a C^∞ function that satisfies the following four conditions:

1. $\lambda \geqq 0$.
2. $\lambda(t) = 0$ when $t \leqq 0$, and $\lambda(t) > 1$ when $t \geqq 1$.
3. $\lambda''(t) \geqq 100\lambda'(t) > 0$ when $t > 0$.
4. $\lambda'(t) > 100$ for t with $\lambda(t) > \dfrac{1}{2}$.

The following domain is called a *worm domain*, due to Diederich and Fornaess:

$$\Omega_{\mathrm{DF},r} := \left\{ z \in \mathbb{C}^2 \middle| \left| z_2 + e^{\sqrt{-1}\log|z_1|^2} \right|^2 - 1 \right.$$
$$\left. + \lambda\left(\frac{1}{|z_1|^2} - 1 \right) + \lambda(|z_1|^2 - r^2) < 0 \right\},$$

where $r > 1$.

$\Omega_{\mathrm{DF},r}$ is a bounded pseudoconvex domain with boundary of class C^∞. If $r \geqq e^\pi$, then

$$\overline{\Omega_{\mathrm{DF},r}} \supset \{(z_1, 0) \,|\, 1 \leqq |z_1| \leqq e^\pi\} \cup \{(z_1, z_2) \,|\, |z_1| = 1 \text{ or } e^\pi, |z_2 + 1| \leqq 1\}.$$

Hence, by an argument similar to the proof of Proposition 3.22, any pseudoconvex domain containing $\overline{\Omega_{\mathrm{DF},r}}$ must contain $\{(z_1, z_2) \mid 1 < |z_1| < e^\pi \text{ and } |z_2 + 1| < 1\}$. In order to construct $\Omega_{\mathrm{DF},r}$, first take a disk on the z_2 plane whose boundary contains the origin, and then let this disk revolve around the origin, changing the center and radius appropriately at the same time the disk travels along the z_1 plane. The point of deriving the pseudoconvexity of $\Omega_{\mathrm{DF},r}$ is that the angle of rotation is a harmonic function of z_1. (For the details, refer to the original article or [**25**].)

REMARK. Recently, a remarkable result on the $\overline{\partial}$ equation on a worm domain has been obtained. Namely, due to M. Christ [**9**], it is known that $P(C_0^\infty(\Omega_{\mathrm{DF},r})) \not\subset C^\infty(\overline{\Omega_{\mathrm{DF},r}})$ for the orthogonal projection

$$P\colon L^2(\Omega_{\mathrm{DF},r}) \longrightarrow A^2(\Omega_{\mathrm{DF},r}).$$

This property of worm domains contrasts finely with the next results.

THEOREM 3.28. *The closure of a pseudoconvex domain whose boundary is of class* C^ω *possesses a system of pseudoconvex neighborhoods.*

For the proof, refer to [**14**].

THEOREM 3.29 (Kohn's theorem). *If* Ω *is a strongly pseudoconvex domain with boundary of class* C^∞, *then* $P(C^\infty(\overline{\Omega})) \subset C^\infty(\overline{\Omega})$, *where* $P\colon L^2(\Omega) \to A^2(\Omega)$ *denotes the orthogonal projection.*

For the proof, refer to [**18**].

As it will be seen in Chapter 6, Theorem 3.29 is useful in studying the boundary behavior of holomorphic mappings.

CHAPTER 4

L^2 Estimates and Existence Theorems

We have already observed that the division and extension problems on $A(\Omega)$ are reduced to the problem of solving the $\overline{\partial}$ equation under appropriate constraints, and have shown that in order for the $\overline{\partial}$ equation to be solvable, Ω must be pseudoconvex. In this chapter, let us go ahead and solve the $\overline{\partial}$ equation on a pseudoconvex open set once without any constraint. The argument stated here follows basically the approaches by J. J. Kohn, L. Hörmander, and others, but the core method of deriving the L^2 estimates is due to the article [**37**]. This modification has practically no effect on the existing results included in this chapter, but will make an essential difference in the next chapter.

In § 4. 1, we derive the vanishing of the $\overline{\partial}$ cohomology on a pseudoconvex open set by means of the L^2 estimates. Here the $\overline{\partial}$ equation is solved under estimates with respect to the L^2 norm with weight function. Fundamental formulae and inequalities are shown in this section. From the vanishing of the $\overline{\partial}$ cohomology, it follows that Serre's condition and pseudoconvexity are equivalent. In § 4. 2, we apply the existence theorem proved in § 4. 1 to generalize the classical results of function theory in one variable to several variables. These results were all established by Kiyoshi Oka, but we follow Hörmander's methods here.

4.1. L^2 Estimates and Vanishing of $\overline{\partial}$ Cohomology

The goal of this section is to show the solvability of the $\overline{\partial}$ equation on a pseudoconvex open set by means of the so-called L^2 estimate. This method is one of the fundamental ideas in functional analysis which originates with the Fredholm alternative, and uses an infinite-dimensional system that keeps the equivalence between the existence of solutions for linear equations and the uniqueness of solutions of their adjoint equations.

Let us review closed operators briefly before starting the details.

Let H_1 and H_2 be Hilbert spaces. A *closed operator* from H_1 to H_2 is by definition a linear mapping T that is defined on a dense linear subset $\mathcal{D}$ in H_1 and has values in H_2 such that the graph

$$G_T := \{(u, Tu) \mid u \in \mathcal{D}\}$$

is a closed set of the direct sum $H_1 \oplus H_2$. $\mathcal{D}$ is called the *domain* of T and denoted by $\operatorname{Dom} T$. Also, $T(\mathcal{D})$ is the *image* of T and denoted by $\operatorname{Im} T$. The denseness of $\operatorname{Dom} T$ determines a closed operator $\widehat{T}$ from H_2 to H_1 whose graph is the orthogonal complement of G_T in $H_1 \oplus H_2$. $-\widehat{T}$ is called the *adjoint operator* of T and denoted by T^*. The definition of T^* may be rephrased in terms of the inner product $(\ ,\ )_i$ of H_i as follows:[1]

$$(4.1)\qquad T^*v = u$$
$$:\Longleftrightarrow (Tw, v)_2 = (w, u)_1 \text{ for any element } w \text{ in } \operatorname{Dom} T.$$

From $(G_T^{\perp})^{\perp} = \overline{G_T} = G_T$, we obtain $(T^*)^* = T$.

In what follows, let $\|\ \|_i$ denote the norm of H_i.

THEOREM 4.1. *The following two statements on an element v in H_2 and a positive number C are equivalent:*

1. *There is an element u in* $\operatorname{Dom} T$ *such that*

$$\begin{cases} Tu = v, \\ \|u\|_1 \leqq C. \end{cases}$$

2. *For any element w in* $\operatorname{Dom} T^*$,

$$|(v, w)_2| \leqq C \,\|T^*w\|_1 .$$

PROOF. $1 \Longrightarrow 2$: Let $w \in \operatorname{Dom} T^*$. From (4.1) it follows that

$$\begin{aligned} |(v, w)_2| &= |(Tu, w)_2| = |(u, T^*w)_1| \\ &\leqq \|u\|_1 \,\|T^*w\|_1 \leqq C \,\|T^*w\|_1 . \end{aligned}$$

$2 \Longrightarrow 1$: Define an antilinear mapping over $\operatorname{Im} T^*$ by

$$\begin{array}{cccc} l: & \operatorname{Im} T^* & \longrightarrow & \mathbb{C} \\ & \cup & & \cup \\ & y = T^*w & \longmapsto & (v, w)_2 . \end{array}$$

From the assumption, $|l(y)| \leqq C \,\|y\|_1$ for any $y \in \operatorname{Im} T^*$. Therefore, from the Hahn–Banach theorem, we obtain an extension $\widetilde{l}$ of l to H_1

[1] $A :\Longleftrightarrow B$ reads 'A is defined by B.' This is a common usage nowadays.

such that

$$|\widetilde{l}(x)| \leqq C\,\|x\|_1 \text{ for any } x \in H_1. \tag{4.2}$$

From the Riesz representation theorem, there exists $u \in H_1$ such that

$$\widetilde{l}(x) = (u, x)_1 \text{ for any element } x \text{ in } H_1. \tag{4.3}$$

In particular, this implies that

$$(v, w)_2 = (u, T^*w)_1 \text{ for any element } w \text{ in } \operatorname{Dom} T^*. \tag{4.4}$$

By applying (4.1) to this statement, we derive $(T^*)^*u = v$, and thus $Tu = v$ since $(T^*)^* = T$. Moreover, it follows from (4.2) and (4.3) that $\|u\|_1 \leqq C$. □

We would like to apply Theorem 4.1 to the $\overline{\partial}$ equation:

$$\overline{\partial}u = v \text{ for } v \in L^{0,\,q}_{\text{loc}}(\Omega) \cap \operatorname{Ker} \overline{\partial} \quad (q > 0)$$

in order to show that there exists a solution u in $L^{0,\,q-1}_{\text{loc}}(\Omega)$ under an appropriate condition.

For this purpose, we consider the Hilbert space

$$L^2_\varphi(\Omega) := \left\{ f \in L^2_{\text{loc}}(\Omega) \;\middle|\; \int_\Omega e^{-\varphi}|f|^2\, dV < \infty \right\}$$

for a real-valued function φ of class C^2 over Ω. Also, let $L^{0,\,q}_\varphi(\Omega)$ be the set of all $(0, q)$-forms with coefficients in $L^2_\varphi(\Omega)$, which inherits the structure of Hilbert space as the direct sum of $L^2_\varphi(\Omega)$'s. Let $u := \sum' u_I\, d\overline{z}_I$ and $v := \sum' v_I\, d\overline{z}_I$ be elements in $L^{0,\,q}_\varphi(\Omega)$. We define the *weighted* L^2 *norm* $\|\ \|_\varphi$ on $L^{0,\,q}_\varphi(\Omega)$ with *weight function* φ by

$$\|u\|^2_\varphi := \sum_I{}' \|u_I\|^2_\varphi, \text{ where } \|u_I\|^2_\varphi := \int_\Omega e^{-\varphi}|u_I|^2\, dV\,. \tag{4.5}$$

(This definition of norms of differential forms is not, in general, compatible with the patching of coordinate neighborhoods in a manifold, but the advantage of fixing a local coordinate system is an enormous simplification of arguments.) Similarly, the *weighted inner product* $(\ ,\ )_\varphi$ on $L^{0,\,q}_\varphi(\Omega)$ with the weight function φ is defined by

$$(u, v)_\varphi := \int_\Omega e^{-\varphi}\langle u, v\rangle\, dV, \text{ where } \langle u, v\rangle := \sum_I{}' u_I \overline{v}_I\,. \tag{4.6}$$

Also, for a multi-index $I := (i_1, \ldots, i_p)$, we set

$$\dot{I} := \{i_1, \ldots, i_p\} \tag{4.7}$$

and

$$I_k^{i_\mu} := (\underbrace{i_1, \ldots, k}_{\mu}, \ldots, i_p), \text{ where } i_\mu \text{ is replaced by } k. \tag{4.8}$$

For $u \in C_0^{0,\,q-1}(\Omega)$ and $v \in C^{0,\,q}(\Omega)$,

$$\begin{aligned}
&\int_\Omega e^{-\varphi} \langle \bar\partial u, v\rangle \, dV \\
&\quad = \int_\Omega e^{-\varphi} {\sum_J}' \sum_{\substack{I \\ \dot I \cup \{k\} = \dot J}}{}' \operatorname{sgn}\binom{J}{kI} \frac{\partial u_I}{\partial \bar z_k} \bar v_J \, dV \\
&\quad = -\int_\Omega e^{-\varphi} {\sum_J}' \sum_{\substack{I \\ \dot I \cup \{k\} = \dot J}}{}' \operatorname{sgn}\binom{J}{kI} u_I \overline{\left(\frac{\partial v_J}{\partial z_k} - \frac{\partial \varphi}{\partial z_k} v_J\right)} \, dV \\
&\quad = -\int_\Omega e^{-\varphi} {\sum_I}' u_I \overline{\left(\sum_{\substack{J \\ \dot I \cup \{k\} = \dot J}}{}' \operatorname{sgn}\binom{J}{kI} \left(\frac{\partial v_J}{\partial z_k} - \frac{\partial \varphi}{\partial z_k} v_J\right)\right)} \, dV.
\end{aligned}$$

Therefore, if we define a differential operator ${}^{\varphi}\vartheta$ of the first order by

$$ {}^{\varphi}\vartheta v := -{\sum_I}' \left(\sum_{\substack{J \\ \dot I \cup \{k\} = \dot J}}{}' \operatorname{sgn}\binom{J}{kI} \left(\frac{\partial v_J}{\partial z_k} - \frac{\partial \varphi}{\partial z_k} v_J\right)\right) d\bar z_I, \tag{4.9}$$

it follows that

$$(\bar\partial u, v)_\varphi = \int_\Omega e^{-\varphi} \langle \bar\partial u, v\rangle \, dV = \int_\Omega e^{-\varphi} \langle u, {}^{\varphi}\vartheta v\rangle \, dV = (u, {}^{\varphi}\vartheta v)_\varphi.$$

Similarly to the case of $\bar\partial$, we extend the domain of ${}^{\varphi}\vartheta$ to $L^{0,\,q}_{\mathrm{loc}}(\Omega)$.

Define an operator ${}^{\varphi}\bar\partial$ from $L_\varphi^{0,\,q-1}(\Omega)$ to $L_\varphi^{0,\,q}(\Omega)$ by

$$\begin{aligned}
&\operatorname{Dom} {}^{\varphi}\bar\partial := \{u \in L_\varphi^{0,\,q-1}(\Omega) \mid \bar\partial u \in L_\varphi^{0,\,q}(\Omega)\} \\
&\quad \Psi \\
&\quad u \longmapsto {}^{\varphi}\bar\partial u := \bar\partial u \in L_\varphi^{0,\,q}(\Omega).
\end{aligned}$$

Then it turns out that ${}^{\varphi}\bar\partial$ is a closed operator[2]. In fact, $\operatorname{Dom} {}^{\varphi}\bar\partial$ is dense in $L_\varphi^{0,\,q-1}(\Omega)$, since clearly $C_0^{0,\,q-1}(\Omega) \subset \operatorname{Dom} {}^{\varphi}\bar\partial$. In order to show that the graph of ${}^{\varphi}\bar\partial$ is closed, it is sufficient to show that $(u, v) \in G_{{}^{\varphi}\bar\partial}$ provided that $(u, v) \in \overline{G_{{}^{\varphi}\bar\partial}}$, i.e. there is a sequence $(u_\nu, {}^{\varphi}\bar\partial u_\nu) \in G_{{}^{\varphi}\bar\partial}$ such that

$$u_\nu \longrightarrow u \text{ and } {}^{\varphi}\bar\partial u_\nu \longrightarrow v \quad (\nu \to \infty).$$

[2]Closed operators of this kind are, in general, called $\bar\partial$ *operators*.

From

$$(u, {}^{\varphi}\vartheta w)_{\varphi} = \lim_{\nu \to \infty} (u_\nu, {}^{\varphi}\vartheta w)_{\varphi}$$
$$= \lim_{\nu \to \infty} (\overline{\partial} u_\nu, w)_{\varphi} = (v, w)_{\varphi},$$

it follows that $\overline{\partial} u = v$. Therefore, $(u, v) = (u, \overline{\partial} u) = (u, {}^{\varphi}\overline{\partial} u)$, which tells us that $\overline{G_{\varphi\overline{\partial}}} = G_{\varphi\overline{\partial}}$.

The adjoint operator of ${}^{\varphi}\overline{\partial}$ is written by ${}^{\varphi}\overline{\partial}^*$. In a similar fashion to ${}^{\varphi}\overline{\partial}$, by restricting ${}^{\varphi}\vartheta$ to the following subset of $L^{0,q}_{\varphi}(\Omega)$:

$$\left\{ v \in L^{0,q}_{\varphi}(\Omega) \mid {}^{\varphi}\vartheta v \in L^{0,q-1}_{\varphi}(\Omega) \right\},$$

we obtain another closed operator, which we denote by the same notation ${}^{\varphi}\vartheta$ for convenience.

In this section, we show that some differential inequality holds by applying integration by parts to an element in $C^{0,q}_0(\Omega)$, and that the range for this inequality to hold, in fact, extends to Dom ${}^{\varphi}\overline{\partial}^* \cap$ Dom ${}^{\varphi}\overline{\partial} \cap L^{0,q}_{\varphi}(\Omega)$ by means of a sort of approximation theorem.

First of all, let us prepare this approximation theorem.

Let $\rho : \Omega \to (0, 1]$ be a C^∞ function such that $\delta_\Omega > \rho \geqq \min\left\{\dfrac{\delta_\Omega}{2}, 1\right\}$ and $\sup_{z \in \Omega} |d\rho(z)| < \infty$.

THEOREM 4.2. *For any element v in* Dom ${}^{\varphi}\overline{\partial} \cap$ Dom ${}^{\varphi}\vartheta \cap L^{0,q}_{\varphi}(\Omega)$, *there exists a sequence $\{v_\mu\}_{\mu=1}^{\infty}$ in $C^{0,q}_0(\Omega)$ such that*

(4.10)
$$\lim_{\mu \to \infty} \left(\|\rho \cdot (\overline{\partial} v_\mu - \overline{\partial} v)\|_{\varphi} + \|\rho \cdot ({}^{\varphi}\vartheta v_\mu - {}^{\varphi}\vartheta v)\|_{\varphi} + \|v_\mu - v\|_{\varphi} \right) = 0.$$

PROOF. Take a C^∞ function $\chi : \mathbb{R} \to \mathbb{R}$ such that $\chi \mid (-\infty, -2) = 1$ and $\chi \mid (-1, \infty) = 0$, and set

$$\chi_R(z) := \chi(-R\rho(z)) \cdot \chi\left(\frac{|z|}{R} - 3\right)$$

for $R > 0$. Then, as $R \to \infty$,

$$\text{(4.11)} \qquad \begin{cases} \rho\,\overline{\partial}(\chi_R v) \longrightarrow \rho\,\overline{\partial} v, \\ \rho\,{}^{\varphi}\vartheta(\chi_R v) \longrightarrow \rho\,{}^{\varphi}\vartheta v, \\ \chi_R v \longrightarrow v, \end{cases}$$

in the sense of convergence with respect to the norm $\| \ \|_{\varphi}$.

In fact, by calculation, we obtain

$$\overline{\partial}(\chi_R v) = \overline{\partial}\chi_R \wedge v + \chi_R \overline{\partial} v.$$

For the second term of the right-hand side, we have

$$\|\rho \cdot (\bar{\partial} v - \chi_R \bar{\partial} v)\|_\varphi \leqq \|\bar{\partial} v - \chi_R \bar{\partial} v\|_\varphi \to 0$$

as $R \to \infty$. As for the first term $\bar{\partial}\chi_R \wedge v$, we have

$$\begin{aligned}\bar{\partial}\chi_R = & -R\chi'(-R\rho(z))\,\chi\left(\frac{|z|}{R} - 3\right)\bar{\partial}\rho \\ & + \frac{\bar{\partial}|z|}{R}\chi(-R\rho(z))\,\chi'\left(\frac{|z|}{R} - 3\right).\end{aligned}$$

Since $\rho \leqq \dfrac{1}{R}$ on $\operatorname{supp}\chi'(-R\rho(z))$, from the boundedness of $|\bar{\partial}\rho|$, it follows that $|\rho\,\bar{\partial}\chi_R \wedge v|$ is bounded when $R \to \infty$. On the other hand, because of the choice of ρ, for any compact set K of Ω, there is a sufficiently large R such that $\bar{\partial}\chi_R \mid K = 0$. Therefore, since $\|\rho\,\bar{\partial}\chi_R \wedge v\|_\varphi \to 0$ as $R \to \infty$, the combination of these results yields $\rho\,\bar{\partial}(\chi_R v) \to \rho\,\bar{\partial} v$.

The other two cases in (4.11) follow similarly.

In order to deduce the desired conclusion from (4.11), it suffices to use the following three kinds of convergence with respect to $\|\ \|_\varphi$:

$$\text{(4.12)} \qquad \begin{cases} \bar{\partial}(\chi_R v)_\varepsilon \longrightarrow \bar{\partial}(\chi_R v), \\ {}^{\varphi}\vartheta(\chi_R v)_\varepsilon \longrightarrow {}^{\varphi}\vartheta(\chi_R v), \\ (\chi_R v)_\varepsilon \longrightarrow \chi_R v \end{cases}$$

as $\varepsilon \to 0$, where $(\chi_R v)_\varepsilon$ is the ε–regularization of $\chi_R v$. This convergence is nothing but a general property of regularizations. (Refer to [28].) □

For the time being, we exhibit several formulae in order to arrange efficiently those terms produced by integration by parts.

For a continuous (a,b)-form ω on Ω, an operator $e(\omega) : L^{p,q}_{\mathrm{loc}}(\Omega) \to L^{p+a,q+b}_{\mathrm{loc}}(\Omega)$ is defined by $e(\omega)(u) := \omega \wedge u$. Let $\iota(\omega)$ denote the adjoint operator of $e(\omega)$, that is, $\iota(\omega)$ is an operator from $L^{p,q}_{\mathrm{loc}}(\Omega)$ to $L^{p-a,q-b}_{\mathrm{loc}}(\Omega)$ such that

$$\langle \omega \wedge u, v\rangle = \langle u, \iota(\omega)(v)\rangle \text{ for } u \in L^{p-a,q-b}_{\mathrm{loc}}(\Omega) \text{ and } v \in L^{p,q}_{\mathrm{loc}}(\Omega).$$

$\iota(\omega)(v)$ is simply written as $\omega \lrcorner v$.

PROPOSITION 4.3. *For* $B := \sum_\alpha B_\alpha d\overline{z}_\alpha \in C^{0,1}(\Omega)$ *and* $u := \sum_I{}' u_I d\overline{z}_I \in C^{0,q}(\Omega)$,

$$B \lrcorner u = \sum_J{}' \sum_\alpha \sum_I{}' \operatorname{sgn}\begin{pmatrix} I \\ \alpha J \end{pmatrix} \overline{B}_\alpha u_I d\overline{z}_J . \tag{4.13}$$

PROOF. For $v = \sum_J{}' v_J d\overline{z}_J \in C^{0,q-1}(\Omega)$,

$$\begin{aligned}
&\langle B \wedge v, u\rangle \\
&= \sum_I{}' \sum_\alpha \sum_J{}' \operatorname{sgn}\begin{pmatrix} I \\ \alpha J \end{pmatrix} B_\alpha v_J \overline{u}_I \\
&= \sum_J{}' v_J \overline{\left(\sum_\alpha \sum_I{}' \overline{B}_\alpha u_I \operatorname{sgn}\begin{pmatrix} I \\ \alpha J \end{pmatrix}\right)} \\
&= \left\langle v, \sum_J{}' \sum_\alpha \sum_I{}' \overline{B}_\alpha u_I \operatorname{sgn}\begin{pmatrix} I \\ \alpha J \end{pmatrix} d\overline{z}_J \right\rangle .
\end{aligned}$$

□

Set $\vartheta := {}^0\vartheta$, and define $\overline{\vartheta}$ by $\overline{\vartheta} u := \overline{(\vartheta \overline{u})}$. Then from (4.9) it is easy to see that

$$\begin{cases} \overline{\partial}\vartheta + \vartheta\overline{\partial} = \partial\overline{\vartheta} + \overline{\vartheta}\partial, \\ \overline{\partial}\,\overline{\vartheta} + \overline{\vartheta}\,\overline{\partial} = 0, \\ \partial\vartheta + \vartheta\partial = 0. \end{cases} \tag{4.14}$$

PROPOSITION 4.4. *For a* C^2 *function* η *on* Ω *and an element* u *in* $C^{n,q}(\Omega)$,

$$\begin{aligned}
&\partial\eta \wedge \overline{\vartheta} u + \overline{\partial}\eta \lrcorner \overline{\partial} u + \overline{\partial}\left(\overline{\partial}\eta \lrcorner u\right) \\
&= \sum_I{}' \sum_{j \in I} \sum_k \frac{\partial^2 \eta}{\partial z_j \partial \overline{z}_k} u_I dz_1 \wedge \cdots \wedge dz_n \wedge d\overline{z}_{I_k^j} \\
&= \sum_I{}' \sum_{j,k} \frac{\partial^2 \eta}{\partial z_j \partial \overline{z}_k} u_{I_j^k} dz_1 \wedge \cdots \wedge dz_n \wedge d\overline{z}_I ,
\end{aligned} \tag{4.15}$$

where we set $u_{I_j^k} = 0$ *for* j *and* k *when* I_j^k *is not defined.*

PROOF. From (4.9),

$$\partial\eta \wedge \overline{\vartheta} u = -\sum_I{}' \sum_j \frac{\partial \eta}{\partial z_j} \frac{\partial u_I}{\partial \overline{z}_j} dz_1 \wedge \cdots \wedge dz_n \wedge d\overline{z}_I . \tag{4.16}$$

From (4.13),

$$(4.17)\ \overline{\partial}\eta \lrcorner \overline{\partial}u = \sum_I{}' \sum_{j\notin \dot{I}} \frac{\partial\eta}{\partial z_j}\frac{\partial u_I}{\partial \overline{z}_j} dz_1\wedge\cdots\wedge dz_n\wedge d\overline{z}_I$$
$$-\sum_I{}' \sum_{j\notin \dot{I}}\sum_{k\in \dot{I}} \frac{\partial\eta}{\partial z_k}\frac{\partial u_I}{\partial \overline{z}_j} dz_1\wedge\cdots\wedge dz_n\wedge d\overline{z}_{I_j^k}.$$

Similarly,

$$(4.18)\ \overline{\partial}(\overline{\partial}\eta \lrcorner u)$$
$$= \overline{\partial}\left(\sum_J{}'\sum_j\sum_I{}' \frac{\partial\eta}{\partial z_j} u_I \operatorname{sgn}\begin{pmatrix} I \\ jJ\end{pmatrix} dz_1\wedge\cdots\wedge dz_n\wedge d\overline{z}_J\right)$$
$$= \sum_I{}'\sum_{j,k} \frac{\partial^2\eta}{\partial z_j\partial \overline{z}_k} u_I dz_1\wedge\cdots\wedge dz_n\wedge d\overline{z}_{I_k^j}$$
$$+\sum_I{}'\sum_{j\in \dot{I}}\sum_k \frac{\partial\eta}{\partial z_j}\frac{\partial u_I}{\partial \overline{z}_k} dz_1\wedge\cdots\wedge dz_n\wedge d\overline{z}_{I_k^j}$$
$$= \sum_I{}'\sum_{j,k} \frac{\partial^2\eta}{\partial z_j\partial \overline{z}_k} u_I dz_1\wedge\cdots\wedge dz_n\wedge d\overline{z}_{I_k^j}$$
$$+\sum_I{}'\sum_{j\in \dot{I}}\sum_{k\notin \dot{I}} \frac{\partial\eta}{\partial z_j}\frac{\partial u_I}{\partial \overline{z}_k} dz_1\wedge\cdots\wedge dz_n\wedge d\overline{z}_{I_k^j}$$
$$+\sum_I{}'\sum_{j\in \dot{I}} \frac{\partial\eta}{\partial z_j}\frac{\partial u_I}{\partial \overline{z}_j} dz_1\wedge\cdots\wedge dz_n\wedge d\overline{z}_I.$$

By adding these formulae, the desired (4.15) is obtained. □

From (4.14) and (4.15) it is easy to derive the following:

PROPOSITION 4.5. *If η is a positive-valued C^2 function on Ω, then*

$$(4.19)\qquad \|\sqrt{\eta}\,\overline{\partial}u\|^2 + \|\sqrt{\eta}\,\vartheta u\|^2$$
$$\geqq 2\operatorname{Re}(\vartheta u, \overline{\partial}\eta\lrcorner u) - \int_\Omega \sum_I{}'\sum_{j,k}\frac{\partial^2\eta}{\partial z_j\partial\overline{z}_k} u_{I_j^k}\overline{u}_I\, dV$$

for $u\in C_0^{n,q}(\Omega)$.

In what follows, (4.19) will be proven again in a more general form for the L^2 norm $\|\ \|_\varphi$ with weight function φ.

For this purpose, define an operator ${}^\varphi\partial$ by

$$(4.20)\qquad {}^\varphi\partial u := \partial u - \partial\varphi\wedge u.$$

Then, for $u \in C_0^{p,q}(\Omega)$ and $v \in C_0^{p+1,q}(\Omega)$,

$$(4.21)\qquad \begin{aligned}({}^{\varphi}\partial u, v)_{\varphi} &= (e^{-\varphi}\cdot {}^{\varphi}\partial u, v)_0\\ &= (\partial(e^{-\varphi}u), v)_0 = (e^{-\varphi}u, \overline{\vartheta} v)_0\\ &= (u, \overline{\vartheta} v)_{\varphi}\,.\end{aligned}$$

Namely, the adjoint operator of ${}^{\varphi}\partial$ with respect to $\| \;\|_{\varphi}$ is equal to $\overline{\vartheta}$ on $C_0^{p,q}(\Omega)$.

On the other hand, for the complex conjugate $\overline{{}^{\varphi}\partial} := \overline{\partial} - e(\overline{\partial}\varphi)$ of ${}^{\varphi}\partial$, we have

$$\begin{aligned}(\overline{{}^{\varphi}\partial} u, v)_0 &= (e^{-\varphi}(\overline{\partial} - e(\overline{\partial}\varphi))u, v)_{-\varphi}\\ &= (\overline{\partial}(e^{-\varphi}u), v)_{-\varphi} = (e^{-\varphi}u, {}^{-\varphi}\vartheta v)_{-\varphi}\\ &= (u, {}^{-\varphi}\vartheta v)_0\,.\end{aligned}$$

Therefore, the adjoint of $\overline{{}^{\varphi}\partial}$ with respect to $\| \;\|_0$ is equal to ${}^{-\varphi}\vartheta$ on $C_0^{p,q}(\Omega)$.

Direct calculation shows that

$$(4.22)\qquad \overline{\vartheta}\,{}^{\varphi}\partial + {}^{\varphi}\partial\overline{\vartheta} = \overline{\vartheta}\partial + \partial\overline{\vartheta} - \overline{\vartheta} e(\partial\varphi) - e(\partial\varphi)\overline{\vartheta}\,.$$

Now take the complex conjugate of both sides of (4.22), consider their adjoint operators, and change the sign of φ so as to obtain

$$(4.23)\qquad {}^{\varphi}\vartheta\overline{\partial} + \overline{\partial}\,{}^{\varphi}\vartheta = \vartheta\overline{\partial} + \overline{\partial}\vartheta + \iota(\overline{\partial}\varphi)\overline{\partial} + \overline{\partial}\iota(\overline{\partial}\varphi).$$

For simplicity, given a C^2 function φ on Ω and a differential form $u = \sum_I{}' u_I dz_1 \wedge \cdots \wedge dz_n \wedge d\overline{z}_I$ or $u = \sum_I{}' u_I d\overline{z}_I$, we define

$$(4.24)\qquad \begin{aligned} L_{\varphi}u &:= \sum_I{}' \sum_{j,k} \frac{\partial^2\varphi}{\partial z_j \partial\overline{z}_k} u_{I_j^k} dz_1 \wedge \cdots \wedge dz_n \wedge d\overline{z}_I\\ &\text{or } \sum_I{}' \sum_{j,k} \frac{\partial^2\varphi}{\partial z_j \partial\overline{z}_k} u_{I_j^k} d\overline{z}_I,\end{aligned}$$

respectively.

Proposition 4.6. *Let φ and ρ be C^2 real-valued functions on Ω. Then, for any element u in $C_0^{n,q}(\Omega)$,*

$$(4.25)\quad \begin{aligned}\|\rho\,\overline{\partial} u\|_{\varphi}^2 + \|\rho\,{}^{\varphi}\vartheta u\|_{\varphi}^2 &= \|\rho\,\overline{\vartheta} u\|_{\varphi}^2 + 4\,\mathrm{Re}\,(\overline{\partial}\rho \lrcorner\, u, \rho\,{}^{\varphi}\vartheta u)_{\varphi}\\ &\quad + (\rho^2 L_{\varphi}u, u)_{\varphi} - (L_{\rho^2}u, u)_{\varphi}\,.\end{aligned}$$

PROOF.

$$\begin{aligned}
&\|\rho\,\overline{\partial}u\|_\varphi^2 + \|\rho\,{}^\varphi\vartheta u\|_\varphi^2 - \|\rho\,\overline{\vartheta}u\|_\varphi^2\\
&= (\rho^2\overline{\partial}u, \overline{\partial}u)_\varphi + ({}^\varphi\vartheta u, \rho^2\,{}^\varphi\vartheta u)_\varphi - (\overline{\vartheta}u, \rho^2\overline{\vartheta}u)_\varphi\\
&= (\rho^2 u, {}^\varphi\vartheta\overline{\partial}u)_\varphi - (\overline{\partial}\rho^2\wedge u, \overline{\partial}u)_\varphi + (u, \rho^2\overline{\partial}\,{}^\varphi\vartheta u)_\varphi\\
&\quad +(u, \overline{\partial}\rho^2\wedge{}^\varphi\vartheta u)_\varphi - (u, \rho^2\,{}^\varphi\partial\overline{\vartheta}u)_\varphi - (u, \partial\rho^2\wedge\overline{\vartheta}u)_\varphi\\
&= (\rho^2 u, \overline{\partial}\varphi\lrcorner\,\overline{\partial}u + \overline{\partial}(\overline{\partial}\varphi\lrcorner\, u) + \partial\varphi\wedge\overline{\vartheta}u)_\varphi\\
&\quad + 2\,\mathrm{Re}\,(u, \overline{\partial}\rho^2\wedge{}^\varphi\vartheta u)_\varphi - (L_{\rho^2}u, u)_\varphi\\
&= 4\,\mathrm{Re}\,(\overline{\partial}\rho\lrcorner\, u, \rho\,{}^\varphi\vartheta u)_\varphi + (\rho^2 L_\varphi u, u)_\varphi - (L_{\rho^2}u, u)_\varphi,
\end{aligned}$$

where we have used both (4.22) and (4.23) for the third equality and (4.16) for the last equality. □

COROLLARY 4.7. *Under the same condition as above, for every positive number C,*

$$\begin{aligned}
(4.26)\qquad &\|\rho\,\overline{\partial}u\|_\varphi^2 + (1+C)\|\rho\,{}^\varphi\vartheta u\|_\varphi^2\\
&\geqq (\rho^2 L_\varphi u, u)_\varphi - (L_{\rho^2}u, u)_\varphi - \frac{2}{C}\|\overline{\partial}\rho\lrcorner\, u\|_\varphi^2\,.
\end{aligned}$$

The calculation passed through (n, q)-forms, but by looking into (4.26) we see that the same result also holds for $(0, q)$-forms. Since we previously set up the $\overline{\partial}$ equation for $(0, q)$-forms, though overlapping a little, Corollary 4.7 can be restated in this form:

PROPOSITION 4.8 (Fundamental inequality). *If φ and ρ are C^2 real-valued functions on Ω, then, for any element u in $C_0^{0,q}(\Omega)$,*

$$\begin{aligned}
(4.27)\qquad &\|\rho\,\overline{\partial}u\|_\varphi^2 + (1+C)\|\rho\,{}^\varphi\vartheta u\|_\varphi^2\\
&\geqq (\rho^2 L_\varphi u, u)_\varphi - (L_{\rho^2}u, u)_\varphi - \frac{2}{C}\|\overline{\partial}\rho\lrcorner\, u\|_\varphi^2\,,
\end{aligned}$$

where C is an arbitrary positive number.

In practice, we have to estimate the right-hand side of (4.27) from below in order to derive the existence theorem from this inequality.

A $(1, 1)$-form $\omega = \sum_{j,k}\omega_{j\overline{k}}dz_j\wedge d\overline{z}_k$ on Ω is said to be *nonnegative* (or *positive*) at a point $x_0\in\Omega$ if the matrix $(\omega_{j\overline{k}}(x_0))$ is a semipositive (or positive) definite Hermitian matrix, respectively. We write $\omega\geqq 0$ (or $\omega > 0$) when ω is nonnegative (or positive), respectively.[3] In

[3] The condition $\omega_1 - \omega_2 \geqq 0$ is written as $\omega_1 \geqq \omega_2$. ($\omega_1 > \omega_2$ is understood similarly.)

terms of this expression, from the definition of L_φ it is clear that $\langle L_\varphi u, u\rangle \geqq 0$ for any $u \in C^{0,q}(\Omega)$ if and only if $\partial\overline{\partial}\varphi \geqq 0$ on Ω.

Also, set

$$\omega \vee u := \sum_I{}' \sum_{j,k} \omega_{j\overline{k}} u_{I^k_j} d\overline{z}_I \tag{4.28}$$

for $u = \sum' u_I d\overline{z}_I$. Then from this definition and (4.13) it is easy to see that

$$\|\overline{\partial}\rho \lrcorner u\|^2_\varphi = \left((\partial\rho \wedge \overline{\partial}\rho) \vee u, u\right)_\varphi . \tag{4.29}$$

Therefore, the fundamental inequality is written as follows:

$$\begin{aligned} &\|\rho\,\overline{\partial}u\|^2_\varphi + (1+C)\|\rho\,{}^\varphi\vartheta u\|^2_\varphi \\ &\geqq \left(\left(\rho^2\partial\overline{\partial}\varphi - \partial\overline{\partial}\rho^2 - \frac{2}{C}\partial\rho \wedge \overline{\partial}\rho\right) \vee u, u\right)_\varphi . \end{aligned} \tag{4.30}$$

From now on, let Ω be pseudoconvex, and fix an unbounded strictly plurisubharmonic exhaustion function ψ of class C^∞ on Ω.

LEMMA 4.9. *Let φ be a strictly plurisubharmonic function of class C^2 on Ω with $L[\varphi] \geqq 1$ everywhere. Then, given any $c \in \mathbb{R}$ and any continuous function $\tau : \mathbb{R} \to \mathbb{R}$, there exists a function $\lambda : \mathbb{R} \to \mathbb{R}$ of class C^2 that satisfies the following four conditions:*

1. *$\lambda(t) = 0$ when $t < c$.*
2. *$\lambda(t) > \tau(t)$ when $t > c+1$.*
3. *$\lambda' \geqq 0$ and $\lambda'' \geqq 0$.*
4. *Letting $\varphi_\lambda := \varphi + \lambda(\psi)$, if $u \in \operatorname{Dom}{}^{\varphi_\lambda}\overline{\partial} \cap \operatorname{Dom}{}^{\varphi_\lambda}\vartheta \cap L^{0,q}_{\varphi_\lambda}(\Omega)$, then*

$$\|\overline{\partial}u\|^2_{\varphi_\lambda} + \|{}^{\varphi_\lambda}\vartheta u\|^2_{\varphi_\lambda} \geqq q\|u\|^2_{\varphi_\lambda}.$$

PROOF. For a natural number ν, we can take a C^∞ function

$$\rho_\nu : \Omega \longrightarrow (0,1]$$

that satisfies both

$$\min\left\{\frac{\nu}{2}\delta_\Omega, 1\right\} \leqq \rho_\nu \leqq \min\{\nu\delta_\Omega, 1\} \tag{4.31}$$

and

$$\left|\frac{\partial\rho_\nu}{\partial z_j}\right| \leqq \nu \quad (j = 1, \cdots, n), \tag{4.32}$$

because of uniform approximation of the continuous function

$$\min\{\nu\delta_\Omega(z), 1\}$$

on Ω by a C^∞ function.

Given $c \in \mathbb{R}$ and choosing ν so that $\rho_\nu = 1$ on $\Omega_{\psi,c}$, we can take λ that satisfies not only (1)–(3) but also the inequality

$$\rho_\nu^2 \partial\overline{\partial}\lambda(\psi) - \partial\overline{\partial}\rho_\nu^2 - 2\nu\partial\rho_\nu \wedge \overline{\partial}\rho_\nu \geqq 0, \tag{4.33}$$

since $\psi \in \mathrm{PSH}^*$.

On the other hand, from the condition $L[\varphi] \geqq 1$, we conclude that $\partial\overline{\partial}\varphi \geqq \partial\overline{\partial}|z|^2$. Therefore, for any element u in $C_0^{0,q}(\Omega)$, (4.30) yields

$$\|\rho_\nu \overline{\partial} u\|_{\varphi_\lambda}^2 + \left(1 + \frac{1}{\nu}\right) \|\rho_\nu {}^{\varphi_\lambda}\vartheta u\|_{\varphi_\lambda}^2 \geqq q\|\rho_\nu u\|_{\varphi_\lambda}^2. \tag{4.34}$$

Since, by Theorem 4.2, (4.34) holds for any element in $\mathrm{Dom}\,{}^{\varphi_\lambda}\overline{\partial} \cap \mathrm{Dom}\,{}^{\varphi_\lambda}\vartheta \cap L_{\varphi_\lambda}^{0,q}(\Omega)$, we obtain (4) by letting $\nu \to \infty$. □

Let us write down what follows immediately from Lemma 4.9 and Theorem 4.1.

PROPOSITION 4.10. *Let Ω and φ_λ be as above. If $w \in \mathrm{Ker}\,\overline{\partial} \cap L_{\varphi_\lambda}^{0,q}(\Omega)$ $(q > 0)$, then there exists an element v in $\mathrm{Dom}\,{}^{\varphi_\lambda}\overline{\partial} \cap L_{\varphi_\lambda}^{0,q-1}(\Omega)$ such that*

$$\begin{cases} \overline{\partial} v = w, \\ q\|v\|_{\varphi_\lambda}^2 \leqq \|w\|_{\varphi_\lambda}^2. \end{cases} \tag{4.35}$$

Likewise, if $w \in \mathrm{Ker}\,{}^{\varphi_\lambda}\vartheta \cap L_{\varphi_\lambda}^{0,q}(\Omega)$ $(q > 0)$, then there exists an element v in $\mathrm{Dom}\,{}^{\varphi_\lambda}\vartheta \cap L_{\varphi_\lambda}^{0,q+1}(\Omega)$ such that

$$\begin{cases} {}^{\varphi_\lambda}\vartheta v = w, \\ q\|v\|_{\varphi_\lambda}^2 \leqq \|w\|_{\varphi_\lambda}^2. \end{cases} \tag{4.36}$$

PROOF. We prove only the first statement, since the second can be done by the same argument. It suffices to show that

$$|(w, u')_{\varphi_\lambda}|^2 \leqq \frac{1}{q}\|w\|_{\varphi_\lambda}^2 \|{}^{\varphi_\lambda}\overline{\partial}^* u'\|_{\varphi_\lambda}^2 \tag{4.37}$$

for any element u' in $\mathrm{Dom}\,{}^{\varphi_\lambda}\overline{\partial}^* \cap L_{\varphi_\lambda}^{0,q}(\Omega)$. Decompose u' as

$$u' = u_1 + u_2, \text{ where } u_1 \in \mathrm{Ker}\,\overline{\partial} \text{ and } u_2 \perp \mathrm{Ker}\,\overline{\partial}.$$

Since $w \in \mathrm{Ker}\,\overline{\partial}$, it follows that

$$(w, u')_{\varphi_\lambda} = (w, u_1)_{\varphi_\lambda}. \tag{4.38}$$

On the other hand, since $u_2 \perp \operatorname{Ker} \overline{\partial}$,

$$v' \in \operatorname{Dom} {}^{\varphi_\lambda}\overline{\partial} \implies 0 = (\overline{\partial} v', u_2)_{\varphi_\lambda} = (v', {}^{\varphi_\lambda}\overline{\partial}^* u_2)_{\varphi_\lambda} . \tag{4.39}$$

Therefore, $u_2 \in \operatorname{Ker} {}^{\varphi_\lambda}\overline{\partial}^*$, which in particular implies that $u_1 \in \operatorname{Dom} {}^{\varphi_\lambda}\overline{\partial}^*$ and ${}^{\varphi_\lambda}\overline{\partial}^* u' = {}^{\varphi_\lambda}\overline{\partial}^* u_1$. Hence, Lemma 4.9 can be applied to u_1 to yield

$$q\|u_1\|^2_{\varphi_\lambda} \leqq \|{}^{\varphi_\lambda}\overline{\partial}^* u_1\|^2_{\varphi_\lambda} = \|{}^{\varphi_\lambda}\overline{\partial}^* u'\|^2_{\varphi_\lambda} . \tag{4.40}$$

Combining this with (4.38), the Cauchy–Schwarz inequality implies (4.37), as we wish. □

The following existence theorem is a fundamental theorem with a wide variety of applications.

THEOREM 4.11 (Hörmander's theorem). *Assume that Ω is pseudoconvex, and a C^2 function $\varphi : \Omega \to \mathbb{R}$ satisfies $L[\varphi] \geqq 1$.*

1. *For any $w \in \operatorname{Ker} \overline{\partial} \cap L^{0,q}_\varphi(\Omega)$ $(q > 0)$, there exists an element v in $L^{0,q-1}_\varphi(\Omega)$ such that $\overline{\partial} v = w$ and $q\|v\|^2_\varphi \leqq \|w\|^2_\varphi$.*
2. *If $w \in \operatorname{Ker} \overline{\partial} \cap L^{0,q}_{\mathrm{loc}}(\Omega)$ $(q < n)$ and the support of w is compact, then there exists an element v in $L^{0,q-1}_{\mathrm{loc}}(\Omega)$ such that the support of v is compact, $\overline{\partial} v = w$, and $(n-q)\|v\|^2_{-\varphi} \leqq \|w\|^2_{-\varphi}$.*

PROOF. (1) For $c \in \mathbb{R}$ and $\tau \equiv 0$, let λ_c denote, expressing the choice of c, a function λ that satisfies the condition of Lemma 4.9. Since λ_c is nonnegative, $\|w\|_{\varphi_{\lambda_c}} \leqq \|w\|_\varphi$ and $w \in L^{0,q}_{\varphi_{\lambda_c}}(\Omega)$.

Therefore, Proposition 4.10 implies that there is some $v_c \in L^{0,q-1}_{\varphi_{\lambda_c}}(\Omega)$ such that

$$\begin{cases} \overline{\partial} v_c = w, \\ q\|v_c\|^2_{\varphi_{\lambda_c}} \leqq \|w\|^2_{\varphi_{\lambda_c}} \ (\leqq \|w\|^2_\varphi) . \end{cases} \tag{4.41}$$

Since $\|v_c\|_{\varphi_{\lambda_c}}$ is bounded with respect to c, $\{v_c\}$ has a subsequence that converges weakly on compact sets. It is sufficient to choose the limit of this subsequence as v.

(2) For $u = \sum' u_I d\overline{z}_I$, set

$$u^* := \sum{}' \operatorname{sgn}\begin{pmatrix} 1\ 2\ \cdots\ n \\ I\,J \end{pmatrix} \overline{u}_J d\overline{z}_J .$$

Then the defining equation (4.9) of ${}^\varphi\vartheta$ is written as

$${}^\varphi\vartheta u = -e^\varphi(\overline{\partial} e^{-\varphi} u^*)^* .$$

Therefore, if $w \in \operatorname{Ker} \overline{\partial}$, then $e^{\varphi_{\lambda_c}} w^* \in \operatorname{Ker} {}^{\varphi_{\lambda_c}}\vartheta$. Also, since the support of w is compact,

$$\|e^{\varphi_{\lambda_c}} w^*\|_{\varphi_{\lambda_c}} = \|e^{\varphi} w^*\|_{\varphi} \ (= \|w\|_{-\varphi})$$

for a sufficiently large c.

Now that the latter half of Proposition 4.10 is applicable to $e^{\varphi_{\lambda_c}} w^*$, there is v_c such that

$$\begin{cases} {}^{\varphi_{\lambda_c}}\vartheta v_c = e^{\varphi_{\lambda_c}} w^*, \\ (n-q)\|v_c\|^2_{\varphi_{\lambda_c}} \leqq \|w\|^2_{-\varphi} \end{cases}$$

for a sufficiently large c.

Apply the above result to $\varphi_{s\lambda_c}$ $(s > 0)$ instead of φ_{λ_c}, and take a sequence of numbers $s_\mu \to \infty$ and a sequence $\{v_\mu\}_{\mu=1}^{\infty}$ that converges with the L^2 norm on compact sets such that

$$\begin{cases} {}^{\varphi_{(\mu)}}\vartheta v_\mu = e^{\varphi_{(\mu)}} w^*, \\ (n-q)\|v_\mu\|^2_{\varphi_{(\mu)}} \leqq \|w\|^2_{-\varphi}, \end{cases}$$

where we set $\varphi_{(\mu)} := \varphi_{s_\mu \lambda_c}$. If we choose, in advance, $\lambda_c(t) > 0$ when $t > c$, then, for $v_\infty := \lim_{\mu \to \infty} v_\mu$,

$$\text{(4.42)} \qquad \begin{cases} \operatorname{supp} v_\infty \subset \overline{\Omega_{\psi,c}}\,, \\ {}^{\varphi}\vartheta v_\infty = e^{\varphi} w^*, \\ (n-q)\|v_\infty\|^2_{\varphi} \leqq \|w\|^2_{-\varphi}\,. \end{cases}$$

Hence, it is enough to put $v := e^{-\varphi} v^*_\infty$. □

Supplement. It is readily seen from the above proof that if

$$\operatorname{supp} w \subset \overline{\Omega_{\psi,c}}\,,$$

then we can take v with $\operatorname{supp} v \subset \overline{\Omega_{\psi,c}}$ as solutions of the $\overline{\partial}$ equation.

As an application of Theorem 4.11, we can derive fundamental results on the representation of $\overline{\partial}$ cohomology groups.

Let us begin by setting up our notation. Define

$$\text{(4.43)} \qquad W^{p,q}_{\mathrm{loc}}(\Omega) := \left\{ u \in L^{p,q}_{\mathrm{loc}}(\Omega) \mid \overline{\partial} u \in L^{p,q+1}_{\mathrm{loc}}(\Omega) \right\}.$$

Then from the complex

$$0 \longrightarrow W^{p,0}_{\mathrm{loc}}(\Omega) \xrightarrow{\overline{\partial}} W^{p,1}_{\mathrm{loc}}(\Omega) \xrightarrow{\overline{\partial}} \cdots \xrightarrow{\overline{\partial}} W^{p,n}_{\mathrm{loc}}(\Omega) \longrightarrow 0\,,$$

the cohomology groups $H^{p,q}_{\mathrm{loc}}(\Omega)$ are determined by

$$\text{(4.44)} \qquad H^{p,q}_{\mathrm{loc}}(\Omega) := \operatorname{Ker} \overline{\partial} \cap W^{p,q}_{\mathrm{loc}}(\Omega) / \{\overline{\partial} u \mid u \in W^{p,q-1}_{\mathrm{loc}}(\Omega)\}\,.$$

If Ω is pseudoconvex, then, given $u \in L^{p,q}_{\mathrm{loc}}(\Omega)$, there is some C^2 function φ with $L[\varphi] \geqq 1$ such that $\|u\|_\varphi < \infty$. Therefore, from Theorem 4.11, we obtain, in particular, a vanishing theorem of cohomology.

THEOREM 4.12. *If Ω is pseudoconvex, then $H^{p,q}_{\mathrm{loc}}(\Omega) = \{0\}$ for $q > 0$.*

From the theorem of L^2 holomorphy, we get

$$H^{p,0}_{\mathrm{loc}}(\Omega) = H^{p,0}(\Omega)\,. \tag{4.45}$$

In effect, this correspondence holds in general.

THEOREM 4.13. *For any open set $\Omega \subset \mathbb{C}^n$, the homomorphisms*

$$\alpha : H^{p,q}(\Omega) \longrightarrow H^{p,q}_{\mathrm{loc}}(\Omega)$$

induced by the injections $C^{p,q}(\Omega) \hookrightarrow L^{p,q}_{\mathrm{loc}}(\Omega)$ are bijections.

PROOF. [4] We can assume that $q > 0$ from the above observation. Also, clearly it is sufficient to prove only the case $p = 0$.

Proof of surjectivity of α. Let $v \in \operatorname{Ker}\bar\partial \cap L^{0,q}_{\mathrm{loc}}(\Omega)$. It is enough to show that there is an element u in $L^{0,q-1}_{\mathrm{loc}}(\Omega)$ such that $v - \bar\partial u \in C^{0,q}(\Omega)$. For this purpose, fix a locally finite covering $\{U_i\}_{i=1}^\infty$ of Ω with $\overline{U_i} \subset \Omega$, where the U_i are open balls; then construct inductively $v_{i_0\cdots i_l} \in L^{0,q-l-1}_{\mathrm{loc}}(U_{i_0} \cap \cdots \cap U_{i_l})$ $(0 \leqq l \leqq q)$ that satisfy the next two conditions:

$$v \mid U_i = \bar\partial v_i\,, \tag{4.46}$$

$$\sum_{\nu=0}^{l} (-1)^\nu v_{i_0\cdots \hat{i}_\nu \cdots i_l} \mid U_{i_0} \cap \cdots \cap U_{i_l} = \bar\partial v_{i_0\cdots i_l}\,, \tag{4.47}$$

where $U_{i_0} \cap \cdots \cap U_{i_l} \neq \emptyset$, and $\hat{i}_\nu$ means the exclusion of the index i_ν. This process is possible since

$$\begin{aligned} &\bar\partial\left(\sum_{\nu=0}^{l}(-1)^\nu v_{i_0\cdots\hat{i}_\nu\cdots i_l}\right) \\ &= \sum_{\nu=0}^{l}\left(\sum_{\mu<\nu}(-1)^{\nu+\mu} v_{i_0\cdots\hat{i}_\mu\cdots\hat{i}_\nu\cdots i_l} + \sum_{\nu<\mu}(-1)^{\nu+\mu+1} v_{i_0\cdots\hat{i}_\nu\cdots\hat{i}_\mu\cdots i_l}\right) \\ &= 0\,, \end{aligned} \tag{4.48}$$

[4] The proof is self-evident if the knowledge of the theory of cohomology with coefficients in sheaves is assumed, but the argument is hands-on, so we give it in detail.

and $H^{0,q}_{\text{loc}}(U_{i_0} \cap \cdots \cap U_{i_l}) = \{0\}$ for $q > 0$ from the pseudoconvexity of $U_{i_0} \cap \cdots \cap U_{i_l}$.

Set

$$u_{i_0 \cdots i_l} := \sum_{\nu=0}^{l} (-1)^{\nu} v_{i_0 \cdots \hat{i}_\nu \cdots i_l} \mid U_{i_0} \cap \cdots \cap U_{i_l} . \tag{4.49}$$

For $l = q$, (4.48) implies that

$$u_{i_0 \cdots i_q} \in \operatorname{Ker} \bar{\partial} \cap L^2_{\text{loc}}(U_{i_0} \cap \cdots \cap U_{i_q}) = A(U_{i_0} \cap \cdots \cap U_{i_q}) . \tag{4.50}$$

Also, (4.49) yields

$$\sum_{\nu=0}^{q+1} (-1)^{\nu} u_{i_0 \cdots \hat{i}_\nu \cdots i_{q+1}} = 0 . \tag{4.51}$$

Use the partition of unity $\{\rho_i\}$ subordinate to $\{U_i\}$ to set

$$u'_{i_0 \cdots i_{q-1}} := \sum_i \rho_i u_{i i_0 \cdots i_{q-1}} ;$$

then, first, by (4.50) we have $u'_{i_0 \cdots i_{q-1}} \in C^{\infty}(U_{i_0} \cap \cdots \cap U_{i_q})$. Secondly,

$$\begin{aligned} \sum_{\nu=0}^{q} (-1)^{\nu} u'_{i_0 \cdots \hat{i}_\nu \cdots i_q} &= \sum_{\nu=0}^{q} \sum_i (-1)^{\nu} \rho_i u_{i i_0 \cdots \hat{i}_\nu \cdots i_q} \\ &= \sum_i \rho_i \sum_{\nu=0}^{q} (-1)^{\nu} u_{i i_0 \cdots \hat{i}_\nu \cdots i_q} \\ &= \sum_i \rho_i u_{i_0 \cdots i_q} \quad \text{by (4.51)} \\ &= u_{i_0 \cdots i_q} . \end{aligned} \tag{4.52}$$

Similarly, for l with $0 \leqq l \leqq q-1$, we can construct $u'_{i_0 \cdots i_{l-1}} \in C^{0,q-l}(U_{i_0} \cap \cdots \cap U_{i_l})$ such that

$$\sum_{\nu=0}^{l+1} (-1)^{\nu} u'_{i_0 \cdots \hat{i}_\nu \cdots i_{l+1}} = \bar{\partial} u'_{i_0 \cdots i_l} . \tag{4.53}$$

From (4.49) and (4.52), it follows that

$$\sum_{\nu=0}^{q} (-1)^{\nu} (v_{i_0 \cdots \hat{i}_\nu \cdots i_q} - u'_{i_0 \cdots \hat{i}_\nu \cdots i_q}) = 0 . \tag{4.54}$$

When $q = 1$, this means that

$$v_i - u'_i = v_j - u'_j . \tag{4.55}$$

When $q > 1$, from (4.54) in the same way that we produced $u'_{i_0\cdots i_{l-1}}$, we can take $v'_{i_0\cdots i_{q-2}} \in W^{0,0}_{\mathrm{loc}}(U_{i_0}\cap\cdots\cap U_{i_{q-2}})$ such that

$$v_{i_0\cdots i_{q-1}} - u'_{i_0\cdots i_{q-1}} = \sum_{\nu=0}^{q-1}(-1)^\nu v'_{i_0\cdots\hat{i}_\nu\cdots i_{q-1}}. \tag{4.56}$$

By applying $\bar\partial$ to both sides of this equation, we obtain

$$\bar\partial v_{i_0\cdots i_{q-1}} - \bar\partial u'_{i_0\cdots i_{q-1}} = \sum_{\nu=0}^{q-1}(-1)^\nu \bar\partial v'_{i_0\cdots\hat{i}_\nu\cdots i_{q-1}}. \tag{4.57}$$

Rewrite this in terms of (4.47); then

$$\sum_{\nu=0}^{q-1}(-1)^\nu(v_{i_0\cdots\hat{i}_\nu\cdots i_{q-1}} - u'_{i_0\cdots\hat{i}_\nu\cdots i_{q-1}} - \bar\partial v'_{i_0\cdots\hat{i}_\nu\cdots i_{q-1}}) = 0. \tag{4.58}$$

Repeat the process of producing (4.58) from (4.54) until we eventually reach the formula

$$v_i - u'_i - \bar\partial v'_i = v_j - u'_j - \bar\partial v'_j. \tag{4.59}$$

Therefore, finally we define an element u in $L^{0,q-1}_{\mathrm{loc}}(\Omega)$ by

$$u := \begin{cases} v_i - u'_i & \text{if } q = 1, \\ v_i - u'_i - \bar\partial v'_i & \text{if } q > 1; \end{cases} \tag{4.60}$$

then this u satisfies

$$v - \bar\partial u\,(= \bar\partial u'_i) \in C^{0,q}(\Omega). \tag{4.61}$$

Hence, the surjectivity of α is proven.

Proof of the injectivity of α. From Corollary 2.10, this is true when $q = 1$. Let us assume that the assertion holds for all k with $1 \leqq k \leqq q-1$. Choose any $w \in \operatorname{Ker}\bar\partial \cap C^{0,q}(\Omega)$ with $w = \bar\partial g$ for some $g \in L^{0,q-1}_{\mathrm{loc}}(\Omega)$. Recalling that $H^{0,l}(\overline{\Delta^n}) = \{0\}$ for $l > 0$, select a similar covering $\{U_i\}$ as above so that for each i there is $s_i \in C^{0,q-1}(U_i)$ satisfying $\bar\partial s_i = w \mid U_i$. Hence, using Theorem 4.12 and the induction hypothesis at the same time, we can inductively construct elements $s_{i_0\cdots i_l}$ in $C^{0,q-l-1}(U_{i_0}\cap\cdots\cap U_{i_l})$ such that

$$\sum_{\nu=0}^{l}(-1)^\nu s_{i_0\cdots\hat{i}_\nu\cdots i_l} \mid U_{i_0}\cap\cdots\cap U_{i_l} = \bar\partial s_{i_0\cdots i_l} \tag{4.62}$$

for $0 \leqq l \leqq q-1$.

Note that $w = \bar{\partial} g$ yields $\bar{\partial}(s_i - g) = 0$. Hence, from Theorem 4.12, there are $t_{i_0 \cdots i_l} \in L^{0,q-l-1}_{\mathrm{loc}}(U_{i_0} \cap \cdots \cap U_{i_l})$ $(1 \leqq l \leqq q-1)$ such that

$$(4.63)\qquad \begin{cases} \bar{\partial} t_i = s_i - g\,, \\ \bar{\partial} t_{i_0 \cdots i_l} = s_{i_0 \cdots i_l} - \sum\limits_{\nu=0}^{l} (-1)^\nu t_{i_0 \cdots \hat{i}_\nu \cdots i_l}\,. \end{cases}$$

In addition, for the case $l = q-1$,

$$(4.64)\qquad \bar{\partial}\left(s_{i_0 \cdots i_{q-1}} - \sum_{\nu=0}^{q-1} (-1)^\nu t_{i_0 \cdots \hat{i}_\nu \cdots i_{q-1}} \right) = 0\,,$$

and so the L^2 holomorphy theorem implies that

$$\sum_{\nu=0}^{q-1} (-1)^\nu t_{i_0 \cdots \hat{i}_\nu \cdots i_{q-1}}$$

is of class C^∞.

Therefore, applying the gluing in terms of the partition of unity used for the proof of surjectivity, first take

$$t'_{i_0 \cdots i_{q-2}} \in C^\infty(U_{i_0} \cap \cdots \cap U_{i_{q-2}})$$

such that

$$\sum_{\nu=0}^{q-1} (-1)^\nu t'_{i_0 \cdots \hat{i}_\nu \cdots i_{q-1}} = \sum_{\nu=0}^{q-1} (-1)^\nu t_{i_0 \cdots \hat{i}_\nu \cdots i_{q-1}}\,,$$

and consecutively choose $t'_{i_0 \cdots i_l} \in C^{0,q-l-2}(U_{i_0} \cap \cdots \cap U_{i_l})$ inductively in the descending order of $0 \leqq l \leqq q-2$ such that

$$\bar{\partial} t'_{i_0 \cdots i_l} = s_{i_0 \cdots i_l} - \sum_{\nu=0}^{l} (-1)^\nu t'_{i_0 \cdots \hat{i}_\nu \cdots i_l}\,.$$

Then eventually we obtain $s_i - \bar{\partial} t'_i = s_j - \bar{\partial} t'_j$. This represents an element of $C^{0,q-1}(\Omega)$, say, v, which results in $\bar{\partial} v = \bar{\partial}(s_i - \bar{\partial} t'_i) = w$. □

From Theorems 4.12 and 4.13, the following vanishing theorem for $\bar{\partial}$ cohomology groups is obtained:

THEOREM 4.14. *If Ω is pseudoconvex, then*

$$H^{p,q}(\Omega) = \{0\} \text{ for } q > 0\,.$$

COROLLARY 4.15. *Pseudoconvex open sets satisfy Serre's condition.*

With this, we have the following implications:

$$\text{Serre's condition} \implies \text{Hartogs pseudoconvexity (Theorem 3.4)},$$
$$\text{Hartogs pseudoconvexity} \iff \text{Pseudoconvexity (Theorem 3.12)},$$
$$\text{Pseudoconvexity} \implies \text{Serre's condition (Theorems 4.12, 4.13)}.$$

Therefore, the condition in Theorem 2.14, for instance, may be replaced by pseudoconvexity. That is to say,

pseudoconvex open sets are domains of holomorphy.

This assertion, named the *Levi problem* (or the *inverse problem of Hartogs*), had been a central conjecture of long standing in the theory of analytic functions of several variables, but was solved by Kiyoshi Oka for the case $n = 2$ in 1942, and by K. Oka [**39**], H. J. Bremermann [**6**], and F. Norguet [**35**] independently for general n.

The first half of Theorem 4.11 has played an active part, but also from the second half we can derive an explicit consequence on the analytic continuation of holomorphic functions.

THEOREM 4.16. *Let K be a bounded closed subset of $\mathbb{C}^n$, and Ω an open set that includes K. If $n \geqq 2$ and if $\Omega \setminus K$ is connected, then the restriction mapping $A(\Omega) \to A(\Omega \setminus K)$ is a surjection.*

PROOF. Fix a neighborhood U of K that is relatively compact in Ω, and let χ be a real-valued function of class C^∞ on Ω with $\chi \mid K = 1$ and $\operatorname{supp} \chi \subset U$. Then, given a holomorphic function f on $\Omega \setminus K$, the trivial extension of $\overline{\partial}((1-\chi)f)$ to $\mathbb{C}^n$ belongs to $C_0^{0,1}(\mathbb{C}^n)$. Take an open ball $\mathbb{B}(0,R)$ for which $\operatorname{supp} \overline{\partial}((1-\chi)f) \subset \mathbb{B}(0,R)$ and $\partial\mathbb{B}(0,R) \cap \Omega \neq \emptyset$. From the latter half of Theorem 4.11, there is a solution u of the equation $\overline{\partial}u = \overline{\partial}((1-\chi)f)$ such that

$$u \in L^2(\mathbb{B}(0,R)) \text{ and } \operatorname{supp} u \Subset \mathbb{B}(0,R). \tag{4.65}$$

In this case, the theorem of L^2 holomorphy implies $(1-\chi)f - u \in A(\Omega)$, while, by the condition $\partial\mathbb{B}(0,R) \cap \Omega \neq \emptyset$ and the theorem of identity, $(1-\chi)f - u$ coincides with f on $\Omega \setminus K$. □

REMARK. Intuitively, K may seem to be extinguished thoroughly by repeating the process of embedding the biholomorphic image of a Hartogs figure into $\Omega \setminus K$ and extending this image to the biholomorphic image of Δ^2. However, it is uncertain whether the function extended by this method is single-valued. In fact, there is an example in which this process cannot be continued without allowing the image of Δ^2 to stick out from Ω in the course. (Refer to [**19**].)

Essentially, the same content as Theorem 4.16 can be described as the extension theorem for functions on a real hypersurface. (The assumption that $n \geqq 2$ is kept valid successively.)

DEFINITION 4.17. Let Ω be a bounded domain whose boundary is of class C^1. A complex-valued function f of class C^1 on $\partial\Omega$ is said to satisfy the *tangential Cauchy–Riemann equation* if there exists an element F in $C^1(\overline{\Omega})$ such that

$$F \mid \partial\Omega = f \text{ and } \overline{\partial} F \wedge \overline{\partial} r \mid \partial\Omega = 0\,,$$

where r is a defining function of Ω.

THEOREM 4.18 (Bochner–Harvey). *Let $\partial\Omega \in C^1$. If an element f in $C^1(\partial\Omega)$ satisfies the tangential Cauchy–Riemann equation, then there exists an element $\widetilde{f}$ in $C^1(\overline{\Omega}) \cap A(\Omega)$ such that $\widetilde{f} \mid \partial\Omega = f$.*

PROOF. We prove this only in the case that $\partial\Omega \in C^\infty$ and $f \in C^\infty(\partial\Omega)$. For the general case, refer to [**22**] and [**27**].

Let $F \in C^\infty(\overline{\Omega})$ with $F \mid \partial\Omega = f$. From the assumption,

$$\overline{\partial} F = \alpha_1 \overline{\partial} r + \beta_1 r \text{ for some } \alpha_1 \in C^\infty(\overline{\Omega}) \text{ and } \beta_1 \in C^{0,1}(\overline{\Omega})\,.$$

Hence, setting $F_1 := F - \alpha_1 r$, we see that

$$F_1 \mid \partial\Omega = f \text{ and } \overline{\partial} F_1 = \beta_1' r, \text{ where } \beta_1' := \beta_1 - \overline{\partial}\alpha_1\,.$$

From $\overline{\partial}(\beta_1' r) = \overline{\partial}(\overline{\partial} F_1) = 0$,

$$\beta_1' \wedge \overline{\partial} r \mid \partial\Omega = 0\,,$$

and so, $\beta_1' = \alpha_2 \overline{\partial} r + \beta_2 r$ for some $\alpha_2 \in C^\infty(\overline{\Omega})$ and $\beta_2 \in C^{0,1}(\overline{\Omega})$. Setting

$$F_2 := F - \alpha_1 r - \frac{\alpha_2}{2} r^2\,,$$

we see that

$$F_2 \mid \partial\Omega = f \text{ and } \overline{\partial} F_2 = \beta_2' r^2, \text{ where } \beta_2' := \beta_2 - \frac{1}{2}\overline{\partial}\alpha_2\,.$$

As the same operation can be repeated, there is a sequence of functions $\{\alpha_k\}_{k=1}^\infty \subset C^\infty(\overline{\Omega})$ such that, given a natural number N,

$$\overline{\partial}\left(F - \sum_{k=1}^N \frac{\alpha_k}{k!} r^k\right) = \beta_N' r^N \text{ for some } \beta_N' \in C^{0,1}(\overline{\Omega})\,.$$

Therefore, there is an element $\widetilde{F}$ in $C^\infty(\overline{\Omega})$ such that $\widetilde{F} \mid \partial\Omega = f$ and the derivatives of $\overline{\partial}\widetilde{F}$ of all orders are equal to 0 on $\partial\Omega$. If we set

$$\omega := \begin{cases} \overline{\partial}\widetilde{F} & \text{on } \overline{\Omega}, \\ 0 & \text{on } \mathbb{C}^n \setminus \overline{\Omega}, \end{cases}$$

then $\omega \in C^{0,1}(\mathbb{C}^n) \cap \operatorname{Ker} \overline{\partial}$ and $\operatorname{supp} \omega \subset \overline{\Omega}$. Hence, the rest of the proof is similar to that of Theorem 4.16. □

REMARK. As to Theorem 4.13, it appears that the proof connects the world of C^∞ functions with that of locally square integrable functions in terms of holomorphic functions. The generality that has developed from arguments of this kind is the so-called theory of cohomology with coefficients in sheaves; and, further, the unobstructed view that has grown by applying this theory to analyzing the singularities of solutions for linear partial differential equations is nothing but the microlocal analysis of M. Sato, T. Kawai, and M. Kashiwara [**40**].

4.2. Three Fundamental Theorems

In the classical general theory there are results that display the perfection of the world of complex functions; here we find the Mittag–Leffler theorem, the Weierstrass theorem and the Runge theorem, and their generalizations to several variables are derived from the established existence theorems, Theorems 4.11–4.14.

4.2.1. Distribution of Poles and Zeros. When a function f is defined on an open set Ω in $\mathbb{C}^n$ except a null set E, f is said to be a *meromorphic function* on Ω if each point $x_0 \in \Omega$ has some neighborhood U $(= U(f, x_0))$ such that on $U \setminus E$, f can be expressed as the quotient of two holomorphic functions defined on U. Let $\mathcal{M}(\Omega)$ denote the set of all meromorphic functions on Ω. Given $f \in \mathcal{M}(\Omega)$, we call

$$f_\infty := \{p \in \Omega \mid \overline{\lim_{z \to p}} |f(z)| = \infty\}$$

the *pole* of f. From the definition, it is obvious that f_∞ is a closed set that is included in E. From Theorem 1.13, f extends over $\Omega \setminus f_\infty$ as a holomorphic function. We identify this extension with f.

DEFINITION 4.19. For an open set U in Ω, the subset

$$(f \mid U \setminus f_\infty) + A(U)$$

of $\mathcal{M}(U)$ is called the *principal part* of f on U and is denoted by $P(f,U)$.

In the case of one variable, the sum of terms of negative power in the Laurent expansion of a meromorphic function at a pole was called the principal part of the function. The above definition is a generalization of this.[5]

The Mittag–Leffler theorem can be generalized to the case of several variables as follows:

THEOREM 4.20. *Let Ω be a pseudoconvex open set, U an open subset of Ω, and $g \in \mathcal{M}(U)$. If g_∞ is a closed subset of Ω, then there exists an element f in $\mathcal{M}(\Omega)$ such that $g \in P(f,U)$.*

PROOF. From the condition, there is a C^∞ function ρ on Ω whose value is 1 on some neighborhood of g_∞ and such that $\operatorname{supp}\rho \subset U$. Set

$$(4.66) \qquad v := \begin{cases} g\,\overline{\partial}\rho & \text{on } U\,, \\ 0 & \text{on } \Omega \setminus U\,. \end{cases}$$

Then $v \in \operatorname{Ker}\overline{\partial} \cap C^{0,1}(\Omega)$. From the pseudoconvexity of Ω, there is a solution $u \in C^\infty(\Omega)$ of the equation $\overline{\partial}u = v$.

Hence, it suffices to define $f := \rho g - u$, where $\rho g \mid \Omega \setminus U = 0$. □

The pole of a meromorphic function is an analytic subset, although we do not prove this fact in the present book.

DEFINITION 4.21. For a holomorphic function f on Ω and an open set $U \subset \Omega$, $f \cdot A(U)$ is called the *divisor class* of f on U and is denoted by $D(f,U)$.

A generalization of the Weierstrass product theorem to several variables is made possible on a pseudoconvex open set whose second Betti number is 0.

THEOREM 4.22. *Let Ω be a pseudoconvex open set, and $H^2(\Omega,\mathbb{Z}) = \{0\}$. If $V(g)$ is a closed set in Ω for an open set U in Ω and $g \in A(U)$, then there exists an element f in $A(\Omega)$ such that $g \in D(f,U)$.*

---

[5]In the above situation, unlike the case of one variable, there does not exist anything that corresponds to the Laurent series. Hence, we are obliged to define this concept as an equivalence class, as in Definition 4.19.

PROOF. Take a locally finite family $\{\mathbb{B}_j\}_{j=1}^{\infty}$ of open balls $\mathbb{B}_j := \mathbb{B}(p_j, R_j)$ in Ω such that

$$\Omega = \bigcup_{j=1}^{\infty} \mathbb{B}_j \tag{4.67}$$

and

$$\mathbb{B}_j \cap V(g) \neq \emptyset \quad \text{implies} \quad \mathbb{B}_j \subset U . \tag{4.68}$$

Define $g_j \in A(\mathbb{B}_j)$ by

$$g_j := \begin{cases} g \mid \mathbb{B}_j & \text{if } \mathbb{B}_j \cap V(g) \neq \emptyset , \\ 1 & \text{if } \mathbb{B}_j \cap V(g) = \emptyset , \end{cases} \tag{4.69}$$

and define $g_{jk} \in A(\mathbb{B}_j \cap \mathbb{B}_k)$ by $g_{jk} := g_j/g_k$, where $\mathbb{B}_j \cap \mathbb{B}_k \neq \emptyset$.

Note that $\mathbb{B}_j \cap \mathbb{B}_k$ is simply connected since it is convex, and that from the definition, g_{jk} does not have any zero point. Hence, we can have some branch u_{jk} of $\log g_{jk}$ be in one-to-one correspondence to (j,k). Then $u_{ijk} := u_{ij} + u_{jk} + u_{ki} \in 2\pi\sqrt{-1}\,\mathbb{Z}$ on $\mathbb{B}_i \cap \mathbb{B}_j \cap \mathbb{B}_k$. Adjusting the u_{jk}'s in advance so that $u_{jk} = -u_{kj}$, we can assume $u_{ijk} + u_{jkl} + u_{kli} + u_{lij} = 0$. Therefore, from the assumption on Ω, there is a set $\{m_{jk}\}$ consisting of elements of $\mathbb{Z}$ such that

$$2\pi\sqrt{-1}(m_{ij} + m_{jk} + m_{ki}) = u_{ij} + u_{jk} + u_{ki} \tag{4.70}$$

on each $\mathbb{B}_i \cap \mathbb{B}_j \cap \mathbb{B}_k$. If we set $\widetilde{u}_{ij} := u_{ij} - 2\pi\sqrt{-1}m_{ij}$, then

$$\widetilde{u}_{ij} + \widetilde{u}_{jk} + \widetilde{u}_{ki} = 0 . \tag{4.71}$$

Let $\{\rho_j\}$ be a partition of unity subordinate to the open covering $\{\mathbb{B}_j\}$, and define

$$u_i := \sum_j \rho_j \widetilde{u}_{ij} , \text{ where } \rho_j \widetilde{u}_{ij} \mid \mathbb{B}_i \setminus \mathbb{B}_j := 0 . \tag{4.72}$$

Then it follows that

$$\begin{aligned} u_i - u_j &= \sum_k \rho_k \widetilde{u}_{ik} - \sum_k \rho_k \widetilde{u}_{jk} \\ &= \sum_k \rho_k (\widetilde{u}_{ik} - \widetilde{u}_{jk}) = \sum_k \rho_k \widetilde{u}_{ij} \\ &= \widetilde{u}_{ij} . \end{aligned} \tag{4.73}$$

Hence, since $\widetilde{u}_{ij} \in A(\mathbb{B}_i \cap \mathbb{B}_j)$, we have $\overline{\partial} u_i = \overline{\partial} u_j$ on $\mathbb{B}_i \cap \mathbb{B}_j$, and so this determines an element in $\operatorname{Ker} \overline{\partial} \cap C^{0,1}(\Omega)$. Therefore, from Theorem 4.14, there is an element u in $C^{\infty}(\Omega)$ such that $\overline{\partial} u = \overline{\partial} u_i$.

If we set $h_i := e^{u_i - u}$, then h_i has no zero point and $h_i \in A(\mathbb{B}_i)$, while from

$$(4.74) \quad \begin{aligned} (u_i - u) - (u_j - u) &= \widetilde{u}_{ij} \\ &\equiv \log(g_i/g_j) \mod 2\pi\sqrt{-1}\,\mathbb{Z} \end{aligned}$$

it follows that $h_i/h_j = g_i/g_j$.

Consequently, if we define $f := g_i/h_i$, then $f \in A(\Omega)$ and $g \in f \cdot A(U)$, which completes the proof. □

4.2.2. Approximation Theorem. According to the Runge approximation theorem in the theory of functions of one variable, a necessary and sufficient condition for the polynomial ring $\mathbb{C}[z]$ to be dense in $A(\Omega)$ for a given open set $\Omega \subset \mathbb{C}$ is that $\mathbb{C} \setminus \Omega$ be connected. This topological condition is related to the theory of functions by the following proposition:

LEMMA 4.23. *A necessary and sufficient condition for $\mathbb{C} \setminus \Omega$ to be connected is that for every compact set K of Ω, there exists a continuous subharmonic exhaustion function $\varphi : \mathbb{C} \to \mathbb{R}$ such that*

$$K \subset \{z \in \Omega \mid \varphi(z) < 0\} \subset \Omega.$$

PROOF. Necessity: By the connectedness of $\mathbb{C} \setminus \Omega$, there is an open set Ω' with a C^∞ boundary such that $K \subset \Omega' \Subset \Omega$ and $\mathbb{C} \setminus \Omega'$ is connected. For this Ω', fix a homeomorphism $\Phi : \mathbb{C} \to \mathbb{C}$ of class C^∞ such that

$$\Phi(\Omega') = \left\{ z \in \mathbb{C} \;\middle|\; \psi(z) := \sum_{\nu=1}^{k} \log|z - \nu| < -R(k) \right\}$$

for some $R(k) \gg 1$, where k denotes the number of connected components of Ω'. Since ψ is subharmonic on $\mathbb{C}$, it does not have any maximal value, and this property is transmitted to $\psi \circ \Phi$. Hence, we can choose an increasing convex function $\lambda : \mathbb{R} \to \mathbb{R}$ (in the broad sense) of class C^∞ that grows so rapidly that $\lambda(\psi \circ \Phi)$ satisfies all the requirements.

Sufficiency: If $\mathbb{C} \setminus \Omega$ were not connected, let E be one of its bounded components. Take a sufficiently large compact set K of Ω such that some bounded component of $\mathbb{C} \setminus K$ contains E, and let φ be such a function as stated in the proposition. Then the set $\{z \mid \varphi(z) \geqq 0\}$ would have a bounded component $\hat{E}\,(\supset E)$. However, from this it follows that φ is subharmonic on the open set $\{z \mid \varphi(z) < 0\} \cup \hat{E}$ while φ has its maximum (greater than or equal to 0) at an interior point. Hence, a contradiction. □

The above interpretation of connectedness is essential to a generalization of the Runge theorem to several variables:

THEOREM 4.24. *For a pseudoconvex open subset Ω of $\mathbb{C}^n$, a necessary and sufficient condition for $\mathbb{C}[z]$ to be dense in $A(\Omega)$ is that for every compact set $K \subset \Omega$, there exists a continuous plurisubharmonic exhaustion function φ defined on $\mathbb{C}^n$ such that $K \subset \{z \mid \varphi(z) < 0\} \subset \Omega$.*

The following generalization of this theorem makes the proof clean-cut.

THEOREM 4.25. *For pseudoconvex open subsets Ω_1 and Ω_2 of $\mathbb{C}^n$ with $\Omega_1 \subset \Omega_2$, a necessary and sufficient condition for $A(\Omega_2)$ to be dense in $A(\Omega_1)$ is that for every compact set K of Ω_1, there exists a continuous plurisubharmonic exhaustion function φ defined on Ω_2 such that $K \subset \{z \mid \varphi(z) < 0\} \subset \Omega_1$.*

PROOF OF SUFFICIENCY. Richberg's theorem enables us to assume that φ is of class C^∞. Let $f \in A(\Omega_1)$. Take a C^∞ function $\chi : \mathbb{R} \to \mathbb{R}$ such that $\chi \,\Big|\, \left(-\infty, \frac{1}{2}\sup_K \varphi\right) = 1$ and $\chi \mid (0, \infty) = 0$, then consider a solution $u \in C^\infty(\Omega_2)$ of the $\bar\partial$ equation $\bar\partial u = \bar\partial(\chi(\varphi)f)$ such that

(4.75)
$$\int_{\Omega_2} e^{-|z|^2 - \lambda(\varphi(z))} |u|^2 \, dV \leqq \int_{\Omega_2} e^{-|z|^2 - \lambda(\varphi(z))} |\bar\partial(\chi(\varphi)f)|^2 \, dV ,$$

where λ is an increasing convex function (in the broad sense) of class C^∞. Since $\varphi > \frac{1}{2}\sup_K \varphi$ on $\operatorname{supp} \bar\partial(\chi(\varphi)f)$, for a given $\varepsilon > 0$ we can take λ, with the condition that $\lambda \,\Big|\, \left(-\infty, \frac{1}{2}\sup_K \varphi\right) = 0$, such that

$$\int_{\Omega_2} e^{-|z|^2 - \lambda(\varphi(z))} |\bar\partial(\chi(\varphi)f)|^2 \, dV < \varepsilon . \tag{4.76}$$

Note that u is holomorphic on $\left\{ z \,\Big|\, \varphi(z) < \frac{1}{2}\sup_K \varphi \right\}$; so, from Cauchy's estimate, there is a constant C that depends only on K and φ such that

$$\sup_K |u| \leqq C \int_{\Omega_2} e^{-|z|^2} |u|^2 \, dV . \tag{4.77}$$

Therefore, defining $\widetilde{f} := \chi(\varphi)f - u$, we see that $\widetilde{f} \in A(\Omega_2)$ and

$$\sup_K |\widetilde{f} - f| = \sup_K |u| < C\varepsilon \,. \tag{4.78}$$

□

Next, we prepare a lemma to show the necessity.

LEMMA 4.26. *If Ω is a strongly pseudoconvex open set, then for any boundary point x_0 of Ω, there exists an element f in $A(\Omega)$ such that*

$$\lim_{z \to x_0} |f(z)| = \infty \,. \tag{4.79}$$

PROOF. From the strong pseudoconvexity, Proposition 3.26 implies that there is a neighborhood $U \ni x_0$ and $g \in A(U)$ such that $V(g) \cap \overline{\Omega} = \{x_0\}$. If we choose a C^∞ function $\rho : \mathbb{C}^n \to [0,1]$ such that $\operatorname{supp} \rho \subset U$ and $\rho \equiv 1$ on a neighborhood of x_0, then since $\overline{\partial}(\rho/g)$ is of class C^∞ on some strongly pseudoconvex neighborhood Ω' of $\overline{\Omega}$, form Theorem 4.14 there is $u \in C^\infty(\Omega')$ such that $\overline{\partial} u = \overline{\partial}(\rho/g)$ on Ω'. In this case, $f := \rho/g - u$ is a holomorphic function on Ω and satisfies (4.79). □

PROOF OF NECESSITY. Let ψ be a strictly plurisubharmonic exhaustion function of class C^∞ on Ω_1, and for a given compact set K in Ω_1, take a real number c such that $K \subset \Omega_{1,c} := \{z \mid \psi(z) < c\}$ and $\Omega_{1,c}$ is strongly pseudoconvex. Then from Lemma 4.26, for each point x_0 of $\partial\Omega_{1,c}$ there is an element f in $A(\Omega_{1,c})$ such that $\lim_{z \to x_0} |f(z)| = \infty$. Hence, the density of $A(\Omega_2)$ in $A(\Omega_1)$, that of $A(\Omega_1)$ in $A(\Omega_{1,c})$, and the compactness of $\partial\Omega_{1,c}$ all together enable us to choose elements $f_1, \cdots, f_m$ in $A(\Omega_2)$ and construct $\widetilde{\varphi}(z) := \sum_{k=1}^m |f_k(z)|^2 - 1$ so that some connected component W of $\Omega_{2,0} := \{z \mid \widetilde{\varphi}(z) < 0\}$ satisfies $K \subset W \subset \Omega_1$.

Therefore, the desired function φ is obtained by setting

$$\varphi := \begin{cases} \widetilde{\varphi} - \dfrac{1}{2} \sup_K \widetilde{\varphi} & \text{on } W \,, \\ \max\left\{ -\dfrac{1}{2} \sup_K \widetilde{\varphi},\ \widetilde{\varphi} - \dfrac{1}{2} \sup_K \widetilde{\varphi} \right\} & \text{on } \Omega_2 \setminus W \,. \end{cases}$$

□

REMARK. A pseudoconvex domain Ω such that $\mathbb{C}[z]$ is dense in $A(\Omega)$ is said to be *polynomially convex.* As the condition for polynomial convexity in one variable was topological, this became a problem in several variables as well, but K. Oka, J. Wermer, and others found counterexamples.

Wermer's counterexample:

1. $K := \{(z, w) \in \mathbb{C}^2 \mid w = \overline{z},\ |\mathrm{Re}\ z| \leqq 1,\ |\mathrm{Im}\, z| \leqq 1\} \Longrightarrow K$ has a fundamental system of neighborhoods consisting of domains U_j that are biholomorphically equivalent to double discs.
2. $\Phi(z, w) := (z,\ (1+\sqrt{-1})w - \sqrt{-1}zw^2 - z^2w^3) \Longrightarrow$ the Jacobian of Φ $(\equiv 1 + \sqrt{-1}) \neq 0$.
3. Φ is one-to-one on a neighborhood of K.
4. $j \gg 1 \Longrightarrow \Phi$ is biholomorphic on U_j.
5. $K \supset \gamma := \{(e^{i\theta}, e^{-i\theta}) \in \mathbb{C}^2 \mid 0 \leqq \theta \leqq 2\pi\}$

$$\begin{aligned} &\Longrightarrow \quad \Phi(\gamma) = \{(e^{i\theta}, 0) \mid 0 \leqq \theta \leqq 2\pi\} \\ &\Longrightarrow \quad \{(z, 0) \in \mathbb{C}^2 \mid |z| \leqq 1\} \\ &\qquad\quad \subset \left\{(z, w) \;\middle|\; |f(z, w)| \leqq \sup_{\gamma}|f|,\ \forall f \in \mathbb{C}[z, w]\right\}. \end{aligned}$$

6. $z \cdot \Phi(z, \overline{z}) = (z^2, |z|^2\{(1-|z|^4)+\sqrt{-1}(1-|z|^2)\}) \Longrightarrow$ the second component of $\Phi(z, \overline{z})$ is not equal to 0 in the range of values $0 < |z| < 1,\ |z| > 1$.
7. From (6), in particular, $\left(\frac{1}{2}, 0\right) \notin \Phi(K) \Longrightarrow \left(\frac{1}{2}, 0\right) \notin \Phi(U_j)$ $(j \gg 1)$.
8. From (5) and (7), $\Phi(U_j)$ is not polynomially convex.

However, the following question remains unsettled.

Bremermann's Problem. Is Ω polynomially convex, provided that for any complex line $l \subset \mathbb{C}^n$, $l \setminus \Omega$ is connected?

The problems of generalizing the Mittag–Leffler theorem and the Weierstrass product theorem to several variables were formulated by Cousin in a more abstract form, and thus they are sometimes called the *first* and *second Cousin problems*, respectively. Oka solved these problems first on domains of holomorphy, and eventually established them as the theory on pseudoconvex open sets by settling the Levi problem.

CHAPTER 5

Solutions of the Extension and Division Problems

In the present chapter, the extension and division problems are solved on pseudoconvex open sets; namely, the $\bar{\partial}$ equation is solved under certain constraints. In § 5. 1, we show how to omit the differentiability from the conditions imposed on weight functions in Theorem 4.11. By this omission, the constraints stemming from the extension problems can be replaced by some conditions on the integrability of relevant weight functions, which produces a general extension theorem. In § 5. 2, the division problem is solved by restricting the domain of the $\bar{\partial}$ operator to an appropriate subspace and inducing an L^2 estimate on this new operator. This approach is due to H. Skoda. § 5. 3 presents an extension theorem which is equipped with a growth rate condition. The content of this theorem asserts that L^2 holomorphic functions defined on the intersection of a hyperplane and Ω extend to Ω under an estimate on the norm which is allowed to involve a plurisubharmonic weight function. § 5. 4 introduces two applications of this L^2 extension theorem; one is a characterization of the removable singularities of L^2 holomorphic functions (Theorem 5.18), and the other the proof of Demailly's approximation theorem (Theorem 3.16).

5.1. Solutions of the Extension Problems

For a general plurisubharmonic function φ on Ω, the Hilbert spaces $L^2_\varphi(\Omega)$ and $L^{p,q}_\varphi(\Omega)$ can be defined as before, because $e^{-\varphi}dV$ is a (Lebesgue) measure on Ω. Noting that $L^2_\varphi(\Omega) \subset L^2_{\text{loc}}(\Omega)$, if the $\bar{\partial}$ equation on such a general $L^{p,q}_\varphi(\Omega)$ is solvable with the norm estimate, then we expect to approach even a deeper structure of $A(\Omega)$ by applying the estimated solutions.

Let us generalize the first half of Theorem 4.11 to this form.

THEOREM 5.1. *Let Ω be a pseudoconvex open set, and $\varphi : \Omega \to [-\infty, \infty)$ a plurisubharmonic function with $L[\varphi] \geqq 1$. Then for any $v \in \operatorname{Ker} \overline{\partial} \cap L^{0,q}_{\varphi}(\Omega)$ $(q > 0)$, there exists an element u in $L^{0,q-1}_{\varphi}(\Omega)$ such that $\overline{\partial} u = v$ and $\|u\|_{\varphi} \leqq \|v\|_{\varphi}$.*

PROOF. Let φ_{ε} be the ε–regularization of φ, and $\Omega_{(\varepsilon)}$ a pseudoconvex open set with $\Omega_{(\varepsilon)} \subset \Omega_{\varepsilon}$. Since $\partial\overline{\partial}\varphi_{\varepsilon} \geqq \partial\overline{\partial}(|z|^2)_{\varepsilon} = \partial\overline{\partial}|z|^2$, Theorem 4.11 is applicable and implies that there is an element u_{ε} in $L^{0,q-1}_{\varphi_{\varepsilon}}(\Omega_{(\varepsilon)})$ such that $\overline{\partial} u_{\varepsilon} = v$ and $\|u_{\varepsilon}\|_{\varphi_{\varepsilon}} \leqq \|v\|_{\varphi_{\varepsilon}}$, where the norm is regarded on $\Omega_{(\varepsilon)}$.

Since $\varphi_{\varepsilon} \geqq \varphi$, $\|v\|_{\varphi_{\varepsilon}} \leqq \|v\|_{\varphi}$. Therefore, from the above estimate for u_{ε}, there is a subfamily of u_{ε} that is weakly convergent on any compact set in Ω. If we denote the limit of this subfamily by u, then $\overline{\partial} u = v$ and $\|u\|_{\varphi} \leqq \|v\|_{\varphi}$. □

Theorem 5.1 is due to Hörmander, but the literature often quotes it in the following form:

COROLLARY 5.2. *Let Ω be a bounded pseudoconvex domain in $\mathbb{C}^n$, and φ a plurisubharmonic function on Ω. Then for any $v \in \operatorname{Ker} \overline{\partial} \cap L^{0,q}_{\varphi}(\Omega)$ $(q > 0)$, there exists an element u in $L^{0,q-1}_{\varphi}(\Omega)$ such that $\overline{\partial} u = v$ and $\|u\|_{\varphi} \leqq C\|v\|_{\varphi}$, where C is a constant that depends only on the diameter of Ω $(:= \sup_{z,z' \in \Omega} |z - z'|)$.*

The interpolation problem raised in Chapter 2 can be solved perfectly as an application of Theorem 5.1.

THEOREM 5.3.[1] *The following are equivalent:*

1. *Ω is pseudoconvex.*
2. *For any discrete set $\Gamma \subset \Omega$, the restriction mapping $A(\Omega) \to \mathbb{C}^{\Gamma}$ is a surjection.*

PROOF. (1) $\Longrightarrow$ (2): Take $h \in C^{\infty}(\Omega)$ such that $\overline{\partial} h = 0$ on some neighborhood of Γ. Then it suffices to show that there is an element g in $L^2_{\text{loc}}(\Omega)$ such that $\overline{\partial} g = \overline{\partial} h$ and $g \mid \Gamma = 0$.

Let ρ and U_{ξ} $(\xi \in \Gamma)$ be as defined in § 2.1, and decompose ρ as

$$\rho = \sum_{\xi \in \Gamma} \rho_{\xi} \text{ with } \operatorname{supp} \rho_{\xi} \subset U_{\xi} . \tag{5.1}$$

[1]One can see also from this theorem that pseudoconvex open sets are domains of holomorphy.

Then for a function Φ defined by

$$\Phi(z) := 2n \sum_{\xi\in\Gamma} \rho_\xi(z) \log |z-\xi|, \tag{5.2}$$

there is a continuous function $\tau : \Omega \to \mathbb{R}$ such that

$$\partial\overline{\partial}\Phi(z) \geqq \tau(z)\partial\overline{\partial}|z|^2 \text{ for } z \in \Omega\backslash\Gamma. \tag{5.3}$$

Hence, if we choose an appropriate exhaustion function ψ on Ω with $\psi \in C^\infty(\Omega) \cap \mathrm{PSH}^*$, then

$$L[\Phi+\psi] \geqq 1 \tag{5.4}$$

and $\|\overline{\partial}h\|_{\Phi+\psi} \leqq 1$.

From Theorem 5.1, there is $g \in L^2_{\Phi+\psi}(\Omega)$ such that $\overline{\partial}g = \overline{\partial}h$ and $\|g\|_{\Phi+\psi} \leqq \|\overline{\partial}h\|_{\Phi+\psi}$. Also, from Theorem 2.7, g is of class C^∞. Noting that $e^{-\Phi-\psi}$ is not integrable around Γ, it follows that $g \mid \Gamma = 0$, which is what we wished to show.

(2) $\Longrightarrow$ (1): This follows from Theorems 3.4 and 3.12. □

Next, coming to the extension of holomorphic functions on an analytic subset X of Ω, two new problems arise if we apply the same argument as in the case of discrete sets:

1. Can holomorphic functions on X be extended to holomorphic functions on some neighborhood of X?
2. Does there exist a function that corresponds to $\Phi+\psi$ in the proof of Theorem 5.3?

As to (1), in general there does not exist any holomorphic mapping from a neighborhood of X to X that coincides with the identity mapping when restricted to X. This problem has already been as difficult as the extension of functions to the whole Ω. Now let us think in a more adaptable way: Given a holomorphic function f on X, construct an extension $\widetilde{f}$ of class C^∞ by patching local holomorphic extensions of f in terms of the partition of unity, and apply Theorem 5.1 to $\overline{\partial}\widetilde{f}$. Then in order to ensure the finiteness of the norm of $\overline{\partial}\widetilde{f}$ by adjusting ψ, $e^{-\Phi}|\overline{\partial}\widetilde{f}|^2$ must be locally integrable in the first place. Since $e^{-\Phi}$ is also required not to be locally integrable along X, it turns out, in turn, that (2) is quite a subtle problem. Precise argument for this point calls for two fundamental theorems on analytic subsets.

THEOREM 5.4 (for the proof, see [**34**] or [**25**]). *Let Ω be an open set in $\mathbb{C}^n$, and X an analytic subset of Ω. Then there exists a family*

$\{X_\alpha\}_{\alpha\in\Lambda}$ of analytic subsets of Ω that satisfies the following conditions:

(5.5) *$X = \bigcup_\alpha X_\alpha$, each X_α is non-empty, and*

$$\#\{X_\alpha \mid X_\alpha \cap K \neq \emptyset\} < \infty$$

for any compact set K of Ω.

(5.6) *Every X_α contains a connected differentiable manifold X'_α as a dense open set in it, and $X'_\alpha \cap X'_\beta = \emptyset$ for $\alpha \neq \beta$.*

An X_α that appears in Theorem 5.4 is called an *irreducible component* of X. Also, the maximum open differentiable manifold contained in X_α is called the *regular part* of X_α and is denoted by $\operatorname{Reg} X_\alpha$. $\operatorname{Reg} X_\alpha$ is a *locally closed complex submanifold* of Ω. Namely, for any point $x \in \operatorname{Reg} X_\alpha$, there are a neighborhood U of x in Ω and a biholomorphic mapping F to Δ^n such that

$$F(U \cap \operatorname{Reg} X_\alpha) = \{z \in \Delta^n \mid z_{m_\alpha+1} = \cdots = z_n = 0\},$$

where m_α is an integer that is independent of the choice of x and is called the *dimension* of X_α. The dimension of X_α is denoted by $\dim X_\alpha$. In addition, $\bigcup_\alpha \operatorname{Reg} X_\alpha$ is called the *regular part* of X and denoted by $\operatorname{Reg} X$.

THEOREM 5.5 (for the proof, see [**34**]). *Let Ω and X be as defined above. Then for any $x \in X$, there exist a neighborhood $U \ni x$ in Ω and a system $\{w_\alpha\}_{\alpha=1}^l$ ($l \neq \infty$) of local defining functions of X on U that possess the following property:*

(5.7) *For any $y \in U \cap X$ and any system $\{h_\beta\}_{\beta=1}^m$ ($1 \leqq m \leqq \infty$) of local defining functions of X around y, there exist a neighborhood V of y in U and a system $\{g_{\alpha\beta}\}_{\alpha=1,\beta=1}^{l,m}$ of holomorphic functions on V such that, for any β,*

$$h_\beta = \sum_{\alpha=1}^{l} w_\alpha g_{\alpha\beta}$$

on V.

From now on, such a system $\{w_\alpha\}$ is called a *reduced system of local defining functions* of X.

THEOREM 5.6. *Holomorphic functions defined on an analytic subset X of a pseudoconvex open set Ω are the restrictions of holomorphic functions on Ω.*

OUTLINE OF THE PROOF. For the reason described above, it suffices to show the existence of a function $\Phi : \Omega \to [-\infty, \infty)$ that satisfies the following conditions:

(5.8) There is some continuous function $\tau : \Omega \to \mathbb{R}$ such that

$$\partial\bar{\partial}\Phi \geqq \tau\partial\bar{\partial}|z|^2 .$$

(5.9) Given a reduced system $\{w_\alpha\}$ of local defining functions of X, the function $e^{-\Phi}\sum|w_\alpha|^2$ is locally square integrable on the domain of w_α.

(5.10) $e^{-\Phi}$ is not integrable around any point of $\operatorname{Reg} X$.

In fact, given a holomorphic function $f : X \to \mathbb{C}$, construct $\widetilde{f}$ by patching local extensions of f. If a Φ that satisfies (5.8)–(5.10) is obtained, then there are elements ψ in $\mathrm{PSH}^* \cap C^\infty(\Omega)$ and u in $L^2_{\Phi+\psi}(\Omega)$ such that $\bar{\partial}u = \bar{\partial}\widetilde{f}$. Since $u \mid X = 0$ from (5.10) and $X = \overline{\operatorname{Reg} X}$, we see that $\widetilde{f} - u$ is the desired extension of f.

Construction of Φ. As it is sufficient to construct Φ for each irreducible component X_α of X, let us assume from the beginning that X is irreducible. Take a locally finite open cover $\{U_j\}$ of Ω so that there is a reduced system $\{w^j_\alpha\}$ of local defining functions of X on each U_j, and define

$$\Phi := (n - m)\log\sum_{j,\alpha}|\rho_j w^j_\alpha|^2 \quad (m := \dim X) \tag{5.11}$$

by means of a partition of unity $\{\rho_j\}$ associated with $\{U_j\}$. Then (5.9) and (5.10) clearly hold. We leave it to the reader to verify (5.8). (Recall the Gauss–Codazzi formula.) □

5.2. Solutions of Division Problems

Given a vector $f = (f_1, \cdots, f_m) \in A(\Omega)^{\oplus m}$ of functions that have no common zero point on Ω, as described in § 2.1, a necessary and sufficient condition for there to exist holomophic functions g_j that satisfy the equation

$$\sum_{j=1}^{m} f_j g_j = 1 \tag{5.12}$$

is that the vector-valued $\bar{\partial}$ equation

$$\bar{\partial}u = v := \left(\bar{\partial}\left(\frac{\bar{f}_1}{|f|^2}\right), \cdots, \bar{\partial}\left(\frac{\bar{f}_m}{|f|^2}\right)\right) \tag{5.13}$$

has a solution $u \in L^2_{\mathrm{loc}}(\Omega)^{\oplus m}$ with $\sum_{j=1}^{m} f_j u_j = 0$.

Let us proceed with the calculation under the assumption that φ is a C^∞ function with $\overline{\partial}(\overline{f}_j/|f|^2) \in L^{0,1}_\varphi(\Omega)$. Define

$$S^{0,q}_\varphi := \left\{ \alpha \in L^{0,q}_\varphi(\Omega)^{\oplus m} \;\middle|\; \sum_{j=1}^{m} f_j \alpha_j = 0 \right\}. \tag{5.14}$$

Then, by the holomorphy of f_j, the $\overline{\partial}$ operator $\alpha \mapsto (\overline{\partial}\alpha_1, \cdots, \overline{\partial}\alpha_m)$ becomes a closed operator from $S^{0,q}_\varphi$ to $S^{0,q+1}_\varphi$, which is denoted by $\overline{\partial}_S$ for distinction.

In order to solve the division problem, it is enough to choose an appropriate φ so that the ratio of $|(w,v)_\varphi|$ to $\|\overline{\partial}^*_S w\|_\varphi$ is bounded on $\operatorname{Ker}\overline{\partial}_S \cap \operatorname{Dom}\overline{\partial}^*_S$.

We will try to express $\overline{\partial}^*_S$ in terms of ${}^\varphi\overline{\partial}^*$ (which operates componentwise) and f. If the problem is solvable, this calculation should naturally produce an L^2 estimate.

Let us first introduce the following notation: For elements w_1 and w_2 in $L^{0,q}_\varphi(\Omega)^{\oplus m}$, set

$$\langle w_1, w_2 \rangle := \sum_{j=1}^{m} \langle w_{1j}, w_{2j} \rangle\,, \tag{5.15}$$

$$(w_1, w_2)_\varphi := \int_\Omega e^{-\varphi} \langle w_1, w_2 \rangle \, dV\,, \tag{5.16}$$

$$\|w_1\|_\varphi := \sqrt{(w_1, w_1)_\varphi}\,. \tag{5.17}$$

The orthogonal complement of the set $S^{0,0}_\varphi$ in $L^{0,0}_\varphi(\Omega)^{\oplus m}$ is denoted by $(S^{0,0}_\varphi)^\perp$. Set

$$(S^{0,0}_\varphi)^\perp_0 := \{ w \mid w = (c\overline{f}_1, \cdots, c\overline{f}_m) \text{ for } c \in C^\infty_0(\Omega) \}. \tag{5.18}$$

Then $(S^{0,0}_\varphi)^\perp_0$ is dense in $(S^{0,0}_\varphi)^\perp$. In fact, since $f_1, \cdots, f_m$ do not have any common zero point, any element in $L^{0,0}_\varphi(\Omega)^{\oplus m}$ that is orthogonal to $S^{0,0}_\varphi$ must be a function-multiple of $(\overline{f}_1, \cdots, \overline{f}_m)$.

Go back to the definition of adjoint operator; then for an element w in $\operatorname{Dom}\overline{\partial}^*_S$ and an element h in $\operatorname{Dom}\overline{\partial}_S$,

$$(\overline{\partial}^*_S w, h)_\varphi = (w, \overline{\partial} h)_\varphi. \tag{5.19}$$

From this, we see that $\overline{\partial}_S^*$ satisfies $\mathrm{Dom}\,{}^{\varphi}\overline{\partial}^* \cap S_{\varphi}^{0,1} \subset \mathrm{Dom}\,\overline{\partial}_S^*$, and that

$$\overline{\partial}_S^* w = P\,{}^{\varphi}\overline{\partial}^* w \tag{5.20}$$

by using the orthogonal projection $P : L_{\varphi}^{0,0}(\Omega)^{\oplus m} \to S_{\varphi}^{0,0}$.

This formula can be expressed as

$$\overline{\partial}_S^* w = {}^{\varphi}\overline{\partial}^* w - \sum_{j=1}^{\infty} ({}^{\varphi}\overline{\partial}^* w, \mathbf{e}_j)_{\varphi}\, \mathbf{e}_j \tag{5.21}$$

in terms of an orthonormal basis $\{\mathbf{e}_j\}_{j=1}^{\infty}$ of $(S_{\varphi}^{0,0})^{\perp}$. We may assume $\mathbf{e}_j \in (S_{\varphi}^{0,0})_0^{\perp}$ in advance. Using the equation $({}^{\varphi}\overline{\partial}^* w, \mathbf{e}_j)_{\varphi} = (w, \overline{\partial}\mathbf{e}_j)_{\varphi}$, the self-duality of Hilbert space allows us to identify $\overline{\partial}\mathbf{e}_j$ with an element in $(L_{\varphi}^{0,1}(\Omega)^{\oplus m})^*$, which results in the expression

$$\overline{\partial}_S^* w = \left({}^{\varphi}\overline{\partial}^* - \sum_{j=1}^{\infty} \mathbf{e}_j \otimes \overline{\partial}\mathbf{e}_j \right) w. \tag{5.22}$$

Since $\mathbf{e}_j = (c_j\overline{f}_1/|f|, \cdots, c_j\overline{f}_m/|f|)$, $c_j \in C_0^{\infty}(\Omega)$, and $\|c_j\|_{\varphi} = 1$, it follows that

$$\overline{\partial}\mathbf{e}_j = \left(\overline{f}_1 \overline{\partial}\left(\frac{c_j}{|f|}\right), \cdots, \overline{f}_m \overline{\partial}\left(\frac{c_j}{|f|}\right) \right) + \left(\frac{c_j}{|f|}\overline{\partial}\overline{f}_1, \cdots, \frac{c_j}{|f|}\overline{\partial}\overline{f}_m \right). \tag{5.23}$$

If $w \in \mathrm{Dom}\,{}^{\varphi}\overline{\partial}^* \cap S_{\varphi}^{0,1}$, the inner product of w with the first term of the right hand side of the above formula is equal to 0. Hence,

$$\begin{aligned} \overline{\partial}_S^* w &= {}^{\varphi}\overline{\partial}^* w - \sum_{j=1}^{\infty}\sum_{k=1}^{m} \left(w_k, \frac{c_j}{|f|}\overline{\partial}\overline{f}_k \right)_{\varphi} \mathbf{e}_j \\ &= {}^{\varphi}\overline{\partial}^* w - \sum_{j=1}^{\infty}\sum_{k=1}^{m} \left(\frac{\overline{\partial}\overline{f}_k}{|f|} \lrcorner\, w_k, c_j \right)_{\varphi} \mathbf{e}_j \\ &= {}^{\varphi}\overline{\partial}^* w - \frac{1}{|f|^2}\left(\sum_{k=1}^{m} \overline{\partial}\overline{f}_k \lrcorner\, w_k \right) \overline{f}. \end{aligned} \tag{5.24}$$

Set for simplicity

$$\beta_f(w) := \frac{\overline{f}}{|f|^2} \sum_{k=1}^{m} \overline{\partial}\overline{f}_k \lrcorner\, w_k.$$

Then from the above formula, for $\rho \in C^{\infty}(\Omega)$ and $\chi \in C_0^{\infty}(\Omega)$,

$$\begin{aligned} &\|\rho\overline{\partial}_S^*(\chi w)\|_{\varphi}^2 \\ &= \|\rho\,{}^{\varphi}\overline{\partial}^*(\chi w)\|_{\varphi}^2 - 2\mathrm{Re}\,(\rho\,{}^{\varphi}\overline{\partial}^*(\chi w),\, \rho\beta_f(\chi w))_{\varphi} + \|\rho\beta_f(\chi w)\|_{\varphi}^2. \end{aligned}$$

Therefore, for any positive number r,

$$\|\rho\bar{\partial}_S^*(\chi w)\|_\varphi^2 \geqq (1-r)\|\rho\,{}^\varphi\bar{\partial}^*(\chi w)\|_\varphi^2 + \left(1-\frac{1}{r}\right)\|\rho\beta_f(\chi w)\|_\varphi^2 .$$

Putting $r := \dfrac{1}{1+\varepsilon}$ $(\varepsilon > 0)$, we get

$$\|\rho\bar{\partial}_S^*(\chi w)\|_\varphi^2 \geqq \frac{\varepsilon}{1+\varepsilon}\|\rho\,{}^\varphi\bar{\partial}^*(\chi w)\|_\varphi^2 - \varepsilon\|\rho\beta_f(\chi w)\|_\varphi^2 .$$

Combine this with the fundamental inequality (4.27); then, when $\rho > 0$, we obtain

$$\begin{aligned}
&\|\rho\bar{\partial}_S^*(\chi w)\|_\varphi^2 + \|\rho\bar{\partial}(\chi w)\|_\varphi^2 \\
&\geqq \frac{\varepsilon}{1+\varepsilon}\|\rho\,{}^\varphi\bar{\partial}^*(\chi w)\|_\varphi^2 + \|\rho\bar{\partial}(\chi w)\|_\varphi^2 - \varepsilon\|\rho\beta_f(\chi w)\|_\varphi^2 \\
&\geqq \frac{\varepsilon}{1+\varepsilon}\cdot\frac{1}{1+C}\left\{((\rho^2 L_\varphi - L_{\rho^2})\chi w, \chi w)_\varphi - \frac{2}{C}\|\bar{\partial}\rho \lrcorner \chi w\|_\varphi^2\right\} \\
&\quad - \varepsilon\|\rho\beta_f(\chi w)\|_\varphi^2 \quad (C > 0).
\end{aligned}$$

Hence, from this point on, as in deriving Theorem 4.11, we obtain the following existence theorem by running ρ and χ by means of the auxiliary weight function φ_λ.

THEOREM 5.7. *Let Ω be a pseudoconvex open set. Assume that elements $f_1, \cdots, f_m$ in $A(\Omega)$ $(m < \infty)$ have no common zero point and that a C^∞ plurisubharmonic function φ on Ω satisfies both*

$$\sum_{k=1}^m \int_\Omega e^{-\varphi}\left|\bar{\partial}\left(\frac{\bar{f}_k}{|f|^2}\right)\right|^2 dV < \infty$$

and $L[\varphi] \geqq (1+\varepsilon)\left\{|f|^{-2}\sum_{k=1}^m |\partial f_k|^2 + 1\right\}$ *for some* $\varepsilon > 0$.

Then there exist elements g_k in $A(\Omega)$ $(k = 1, \cdots, m)$ such that $\sum_{k=1}^m f_k g_k = 1$, *and*

$$\int_\Omega e^{-\varphi}|g|^2\, dV \leqq \int_\Omega e^{-\varphi}\left(1+\frac{1}{\varepsilon}\right)\sum_{k=1}^m \left|\bar{\partial}\left(\frac{\bar{f}_k}{|f|^2}\right)\right|^2 dV.$$

COROLLARY 5.8. *If Ω is pseudoconvex, then for any system $\{f_k\}_{k=1}^\infty$ of holomorphic functions on Ω that has no common zero point, there exists a system $\{g_k\}_{k=1}^\infty$ of holomorphic functions on Ω such that $\sum_{k=1}^\infty f_k g_k = 1$. In particular,* $\operatorname{Spec}_m A(\Omega) = \Omega$.

PROOF. We may assume that $0 < \sum_{k=1}^{\infty} |f_k|^2 < \infty$. Then from Cauchy's estimate it follows that $\sum_{k=1}^{\infty} |\partial f_k|^2 \in C^\infty(\Omega)$. Hence, there is a plurisubharmonic exhaustion function φ of class C^∞ on Ω such that

$$\begin{cases} \sum_{k=1}^{\infty} \int_\Omega e^{-\varphi} \left| \bar{\partial} \left(\frac{\overline{f}_k}{|f|^2} \right) \right|^2 dV < \infty , \\ L[\varphi] \geqq 2 \left(|f|^{-2} \sum_{k=1}^{\infty} |\partial f_k|^2 + 1 \right) . \end{cases}$$

For any $c < \sup \varphi$, we can take a sufficiently large integer $m = m_c$ so that $f_1, \cdots, f_m$ have no common zero point in $\Omega_{\varphi,c}$. From Theorem 5.7, there are holomorphic functions $g_{c,k}$ on $\Omega_{\varphi,c}$ $(k = 1, \cdots, m)$ such that

$$\begin{cases} \sum_{k=1}^{m} f_k g_{c,k} = 1, \\ \int_\Omega e^{-\varphi} \sum_{k=1}^{m} |g_{c,k}|^2 dV \leqq M, \end{cases}$$

where M is a constant that does not depend on c or m.

Therefore, from Cauchy's estimate and Montel's theorem, there is a sequence of numbers c_μ with $c_\mu \nearrow \sup \varphi$ such that the sequence $\{g_{c_\mu,k}\}_{\mu=1}^{\infty}$ of functions is uniformly convergent on compact subsets of Ω for each k. Define

$$g_k := \lim_{\mu \to \infty} g_{c_\mu,k} ;$$

then from Weierstrass' double series theorem it follows that $g_k \in A(\Omega)$ and

$$\sum_{k=1}^{\infty} f_k g_k = 1.$$

□

If $\mathrm{Spec}_m A(\Omega) = \Omega$, then in particular, since for any point a of $\partial\Omega$ there are elements g_j in $A(\Omega)$ $(j = 1, \cdots, n)$ such that

$$\sum_{j=1}^{n} (z_j - a_j) g_j(z) = 1,$$

it follows that Ω is a domain of holomorphy, and thus pseudoconvex due to Oka's theorem. Hence, the combination of this with Corollary 5.8 shows the converse of Proposition 2.1.

The condition of Theorem 5.7 involves the derivatives of f_k. The above argument does not especially have anything artificial, and this is good enough. However, the result will be neatly stated if we find the calculation described below.

Let v be as defined in (5.13). For $w \in S_\varphi^{0,1} \cap C_0^{0,1}(\Omega)^{\oplus m}$, it is desirable to evaluate $|(w,v)_\varphi|$ by a constant multiple of $\|\overline{\partial}_S^* w\|_\varphi + \|\overline{\partial} w\|_\varphi$ from above; but since

$$\begin{aligned}(w,v)_\varphi &= (w,\overline{\partial}u)_\varphi \\ &= ({}^\varphi\overline{\partial}^* w, u)_\varphi \\ &= (\overline{\partial}_S^* w, u)_\varphi + (\beta_f(w), u)_\varphi,\end{aligned}$$

it suffices to evaluate $\|\beta_f(w)\|_\varphi$. What is readily seen from the form of $\beta_f(w)$ is that if we take $\varphi + p\log|f|^2$ as a weight function instead of φ, then $\|\beta_f(w)\|^2_{\varphi+p\log|f|^2}$ is absorbed in $(L_{\varphi+p\log|f|^2}w, w)_{\varphi+p\log|f|^2}$ when $p \gg 1$. Afterwards, accurate evaluation of the quadratic form implies the following theorem:

THEOREM (Skoda's Theorem [**45**]). *Let Ω be a pseudoconvex open set, and φ a plurisubharmonic function on Ω. Suppose that we are given p holomorphic functions $g_1, \cdots, g_p$ (or a sequence $\{g_j\}_{j=1}^\infty$ of holomorphic functions) on Ω. Let $\alpha > 1$, and $q := \inf\{n, p-1\}$ (or $q := n$). If a holomorphic function f on Ω satisfies*

$$\int_\Omega |f|^2|g|^{-2\alpha q-2}e^{-\varphi}dV < \infty,$$

then there exist p holomorphic functions h_j (or there exists a sequence $\{h_j\}_{j=1}^\infty$ of holomorphic functions) on Ω such that

$$\begin{cases} f = \sum_{j=1}^{p} g_j h_j \quad \left(\text{or } f = \sum_{j=1}^{\infty} g_j h_j\right), \text{ and} \\ \int_\Omega |h|^2|g|^{-2\alpha q}e^{-\varphi}dV \leqq \dfrac{\alpha}{\alpha-1}\int_\Omega |f|^2|g|^{-2\alpha q-2}e^{-\varphi}dV, \end{cases}$$

respectively, where $\sum_{j=1}^{\infty} g_j h_j$ is the sum in the sense of the uniform convergence on compact subsets in Ω.

5.3. Extension Theorem with Growth Rate Condition

5.3.1. L^2 Extension Theorem. In what follows Ω is assumed to be a pseudoconvex open set.

As in § 4. 1, for a general plurisubharmonic function φ on Ω, consider a Hilbert space $L^2_\varphi(\Omega)$, and set $A^2_\varphi(\Omega) := A(\Omega) \cap L^2_\varphi(\Omega)$. $A^2_\varphi(\Omega)$ is a closed subspace of $L^2_\varphi(\Omega)$ due to Cauchy's estimate.

Put $H := \{z \in \mathbb{C}^n \mid z_n = 0\}$ and $\Omega' := \Omega \cap H$. Then consider

$$A^2_\varphi(\Omega') = \left\{ f \in A(\Omega') \;\middle|\; \int_{\Omega'} e^{-\varphi} |f|^2 dV < \infty \right\}$$

as a subspace of $A(\Omega')$. From Theorem 5.5 (or Theorem 2.5 + Theorem 4.14), there is a mapping

$$I : A^2_\varphi(\Omega') \longrightarrow A(\Omega)$$

such that $I(f) \mid \Omega' = f$ for every $f \in A^2_\varphi(\Omega')$.

The problem arising here is about the existence of such an I that is also a bounded linear mapping from $A^2_\varphi(\Omega')$ to some subspace $A^2_\psi(\Omega)$ of $A(\Omega)$. When this condition is satisfied, I is said to be an *interpolation operator* from $A^2_\varphi(\Omega')$ to $A^2_\psi(\Omega)$. Of course, it depends on the relation between φ and ψ whether or not there is an interpolation operator.

Before interpreting this into the problem of $\overline{\partial}$ equation, let us reduce the situation to a specific case.

Take an increasing sequence $\{\Omega_k\}_{k=1}^\infty$ of strongly pseudoconvex open sets of Ω such that $\Omega_k \Subset \Omega_{k+1}$ and $\bigcup_{k=1}^\infty \Omega_k = \Omega$. Then set $\Omega'_k := \Omega_k \cap H$. Also, given two plurisubharmonic functions φ_1 and φ_2 on Ω, let $\varphi_{i,\varepsilon}$ be the ε–regularization of φ_i.

The following will be self-evident:

PROPOSITION 5.9. *Let $\{\Omega_k\}_{k=1}^\infty$ be as above. If for some sequence of positive numbers ε_k that converge to 0 there exist interpolation operators*

$$I_k : A^2_{\varphi_{2,\varepsilon_k}}(\Omega'_{k+1}) \longrightarrow A^2_{\varphi_{1,\varepsilon_k}}(\Omega_k)$$

whose norms form a bounded sequence on k, then there exists an interpolation operator $I : A^2_{\varphi_2}(\Omega') \to A^2_{\varphi_1}(\Omega)$ whose norm does not exceed $\varlimsup_{k\to\infty} \|I_k\|$, where $\|I_k\|$ denotes the norm of I_k.

Therefore, in order to finish this argument, it suffices to construct a linear mapping $I : A^2_{\varphi_2}(\Omega'_2) \to A^2_{\varphi_1}(\Omega_1)$ that satisfies $I(f) \mid \Omega'_1 =$

$f \mid \Omega_1'$ with a certain estimate on the norm for two given strongly pseudoconvex domains $\Omega_1 \Subset \Omega_2$ and $\varphi_1, \varphi_2 \in \mathrm{PSH} \cap C^\infty(\Omega_2)$.

Let Ω_i and φ_i $(i = 1, 2)$ be as above, and take $f \in A^2_{\varphi_2}(\Omega_2')$. The $\overline{\partial}$ equations that are necessary in the procedure for construction of I are obtained as follows:

By means of the projection

$$\begin{array}{rccc} p: & \mathbb{C}^n & \longrightarrow & H \\ & \Cup & & \Cup \\ & z & \longmapsto & z' := (z_1, \cdots, z_{n-1}), \end{array}$$

we extend $f = f(z')$ to a function $p^* f(z) := f(p(z))$ on $p^{-1}(\Omega_2')$.

Choose a positive number δ such that

$$p^{-1}(\Omega_2') \cap \{z \mid |z_n| < \delta\} \supset \Omega_{1,\delta} := \Omega_1 \cap \{z \mid |z_n| < \delta\}.$$

Take a C^∞ function $\chi : \mathbb{R} \to [0, 1]$ that satisfies

$$\chi(t) = \begin{cases} 1 & \text{for } t < \dfrac{1}{2}\,, \\ 0 & \text{for } t > 1\,, \end{cases} \tag{5.25}$$

and for $\chi_\delta(z) := \chi\left(\dfrac{|z_n|}{\delta}\right)$, set

$$v_\delta := \begin{cases} p^* f \cdot \overline{\partial} \chi_\delta & \text{for } z \in \Omega_{1,\delta}\,, \\ 0 & \text{for } z \in \Omega_1 \backslash \Omega_{1,\delta}\,. \end{cases}$$

Then it follows that $v_\delta \in \operatorname{Ker} \overline{\partial} \cap C^{0,1}(\Omega_1)$ and

$$\operatorname{supp} v_\delta \subset \left\{ z \;\middle|\; \frac{\delta}{2} < |z_n| < \delta \right\}.$$

At this point, if there is a $u_\delta \in L^2_{\varphi_1}(\Omega_1)$ such that

$$\begin{cases} \overline{\partial} u_\delta = v_\delta, \\ \|u_\delta\|_{\varphi_1} \leqq C \|v_\delta\|_{\varphi_2} \\ \text{for some constant } C \text{ independent of } f \text{ and } \delta \text{, and} \\ \dfrac{u_\delta}{z_n} \in L^2_{\mathrm{loc}}(\Omega_1), \end{cases} \tag{5.26}$$

and if the correspondence $f \mapsto u_\delta$ can be made linear, then for a sufficiently small δ, the linear mapping

$$\begin{array}{rccc} I_\delta : & A^2_{\varphi_2}(\Omega_2') & \longrightarrow & A^2_{\varphi_1}(\Omega_1) \\ & \Cup & & \Cup \\ & f & \longmapsto & p^* f \cdot \chi_\delta - u_\delta \end{array}$$

will obviously satisfy

$$\begin{cases} I_\delta(f) \mid \Omega_1' = f \mid \Omega_1', \\ \|I_\delta(f)\|_{\varphi_1}^2 \leqq (C+1) \int_{\Omega_2} e^{-\varphi_2} |f|^2 dV_{n-1}. \end{cases}$$

This argument has clarified what kind of $\overline{\partial}$ equation should be treated, but here we will solve the equation only for the case $\varphi_1 = \varphi_2$ for simplicity, and will refer the reader to the literature for the general case.

The next theorem is contained in the author's joint paper [**37**] with K. Takegoshi. The proof in the original article was written in the framework of differential geometry as a development of the Kodaira–Nakano vanishing theorem, but as described below, the proof can be done without displaying the concepts of metric and curvature for some sequence of positive numbers ε_k that converge to 0. (A similar approach can be seen in [**3**] and [**44**].)

THEOREM 5.10 (L^2 extension theorem). *Given a plurisubharmonic function φ on a bounded pseudoconvex open set Ω, there exists an interpolation operator from $A^2_\varphi(\Omega')$ to $A^2_\varphi(\Omega)$ whose norm does not exceed a constant that depends only on the diameter of Ω.*

PROOF. Recall the result of Proposition 4.6, that for any element u in $C_0^{n,1}(\Omega)$,

$$(5.27)\quad \|\rho\overline{\partial}u\|_\Phi^2 + \|\rho\, {}^\Phi\vartheta u\|_\Phi^2 = \|\rho\overline{\vartheta}u\|_\Phi^2 + 4\mathrm{Re}\,(\overline{\partial}\rho \lrcorner u, \rho\, {}^\Phi\vartheta u)_\Phi + (\rho^2 L_\Phi u, u)_\Phi - (L_{\rho^2}u, u)_\Phi,$$

where ρ and Φ are arbitrary real-valued functions of class C^2 on Ω. We assume $\rho > 0$ below.

As the 'error term' $\|\overline{\partial}\rho \lrcorner u\|^2$ in the fundamental inequality is not easy to evaluate in this case, we will use, instead, the following inequality:

$$(5.28)\quad \begin{aligned} &\|\rho\sqrt{\rho^3+1}\, {}^\Phi\vartheta u\|_\Phi^2 + \|\rho\overline{\partial}u\|_\Phi^2 \\ &\geqq ((\rho^2 L_\Phi - L_{\rho^2})u, u)_\Phi - 4\|\rho^{-2}\overline{\partial}\rho \lrcorner u\|_\Phi^2 \\ &=: Q_{\rho,\Phi}(u), \end{aligned}$$

which is obtained by modifying (5.27) in terms of

$$|4\mathrm{Re}\,(\overline{\partial}\rho \lrcorner u, \rho\, {}^\Phi\vartheta u)_\Phi| \leqq \|\rho^3\, {}^\Phi\vartheta u\|_\Phi^2 + 4\|\rho^{-2}\overline{\partial}\rho \lrcorner u\|_\Phi^2 .$$

It is obvious that (5.28) can apply to an element in $C_0^{0,q}(\Omega)$.

Let us consider the $\overline{\partial}$ equations (5.26) under the assumption that $\Omega_1 \Subset \Omega_2 \subset \Omega$ and $\varphi_1 = \varphi_2 \in \mathrm{PSH} \cap C^\infty(\Omega_2)$.

Since ρ is contained as a factor in both norms on the left hand side of (5.28), as in deriving Theorem 4.11, the approximation principle based on Theorem 4.2 implies the following: If there exists a constant C'_Ω such that for any $w \in C_0^{0,1}(\Omega_1)$ we have

$$\left|\left(\frac{v_\delta}{z_n}, w\right)_\Phi\right|^2 \leqq C'_\Omega \, Q_{\rho,\Phi}(w) \int_{\Omega'_2} e^{-\varphi_1} |f|^2 \, dV_{n-1} , \tag{5.29}$$

then there exists a unique element $u'_\delta \in L^2_\Phi(\Omega_1)$ that satisfies

$$\overline{\partial}(\rho\sqrt{\rho^3+1}\, u'_\delta) = \frac{v_\delta}{z_n} , \tag{5.30}$$

$$\|u'_\delta\|^2_\Phi \leqq C'_\Omega \int_{\Omega'_2} e^{-\varphi_1} |f|^2 dV_{n-1}, \tag{5.31}$$

$$u'_\delta \perp \mathrm{Ker}\,(\overline{\partial} \circ \rho\sqrt{\rho^3+1}) . \tag{5.32}$$

If the diameter of Ω is denoted by d_Ω, then

$$\eta(z) := -\log|z_n|^2 + 2\log d_\Omega > 0 \quad (z \in \Omega) .$$

Hence, putting $\eta_\varepsilon(z) := -\log(|z_n|^2 + \varepsilon^2) + 2\log d_\Omega + 3$, if the positive number ε is sufficiently small, then $\eta_\varepsilon > 2$ on Ω.

In order for (5.29) to hold, we first take δ so that $\eta_\delta > 2$ in advance, and then set

$$\rho := \sqrt{\eta_\delta + \log \eta_\delta} ,$$
$$\Phi := \varphi_1 .$$

In this case, since

$$\begin{aligned} Q_{\rho,\Phi}(w) \;&\geqq\; (\partial\overline{\partial}(\log(|z_n|^2+\delta^2) - \log\eta_\delta) \vee w,\, w)_{\varphi_1} \\ &\quad - \frac{1}{4}\|\eta_\delta^{-1}\partial(\eta_\delta + \log\eta_\delta) \lrcorner\, w\|^2_{\varphi_1} \\ &\geqq\; (\partial\overline{\partial}\log(|z_n|^2+\delta^2) \vee w,\, w)_{\varphi_1} \\ &\geqq\; \int_{\{\frac{1}{2}<|z_n|<1\}\cap\Omega_1} e^{-\varphi_1} \frac{\delta^2}{(|z_n|^2+\delta^2)^2} |w_n|^2 \, dV \\ &\quad \left(\text{where } w := \sum_{j=1}^n w_j d\overline{z}_j\right) , \end{aligned} \tag{5.33}$$

the Cauchy–Schwarz inequality implies

$$(5.34)\quad \left|\left(\frac{v_\delta}{z_n}, w\right)_{\varphi_1}\right|^2 \leqq \int_{\Omega_1} e^{-\varphi_1}\frac{(|z_n|^2+\delta^2)^2}{\delta^2}\frac{|v_\delta|^2}{|z_n|^2}\,dV \cdot \int_{\{\frac{1}{2}<|z_n|<1\}\cap\Omega_1} e^{-\varphi_1}\frac{\delta^2}{(|z_n|^2+\delta^2)^2}|w_n|^2\,dV$$

$$\leqq \frac{25}{4}\int_{\Omega_1} e^{-\varphi_1}|v_\delta|^2\,dV\cdot Q_{\rho,\Phi}(w)\,.$$

Therefore, if we take $C'_\Omega = \dfrac{25}{4}\pi \sup|\chi'|^2$, then (5.29) holds when δ is sufficiently small.

In this case, for u'_δ that satisfies (5.30)–(5.32), if we set

$$u_\delta := z_n u'_\delta\,,$$

then it follows that $\bar\partial u_\delta = v_\delta$, $z_n^{-1}u_\delta \in L^2_{\mathrm{loc}}(\Omega_1)$, and

$$\begin{aligned}(5.35)\quad \|u_\delta\|_{\varphi_1} &= \|z_n\,\rho\sqrt{\rho^3+1}\,u'_\delta\|_{\varphi_1}\\ &\leqq \sup_{z\in\Omega_1}|z_n|\,\rho\sqrt{\rho^3+1}\,\|u'_\delta\|_{\varphi_1}\\ &\leqq \sup_{0<t<d_\Omega}\sqrt{t}\,(-\log t+\log(-\log t)+1)^{5/2}\|u'_\delta\|_{\varphi_1}\\ &\leqq C''_\Omega\|u'_\delta\|_{\varphi_1}\,,\end{aligned}$$

where C''_Ω is a constant that depends only on d_Ω.

Combination of (5.31) with (5.35) shows that u_δ is determined in the way of being linearly dependent on f and satisfies (5.26) for $C := \sqrt{C'_\Omega}\,C''_\Omega$, from which Proposition 5.9 yields the desired conclusion. □

5.3.2. Generalizations of the L^2 Extension Theorem. In Theorem 5.10, it is clear that the boundedness of Ω can be replaced by $\sup_\Omega |z_n| < \infty$. When Ω is a general unbounded pseudoconvex open set, there may or may not exist an interpolation operator depending on the condition about the weight function φ. We will give a class of weight functions that guarantee the existence of interpolation operators.

First we consider

$$\mathcal{G}_\Omega := \{G : \Omega \to [-\infty, 0) \mid G \text{ is continuous, and } G - \log|z_n|^2 \in C^2(\Omega)\}$$

as an auxiliary family of functions, and

$$\Pi_\kappa(\Omega) := \{\varphi \in \mathrm{PSH}(\Omega) \mid \varphi + G,\ \varphi + \kappa G \in \mathrm{PSH}(\Omega) \text{ for some } G \in \mathcal{G}_\Omega\}$$

as a practical class of weight functions. Set $l_G := (G - \log|z_n|^2) \mid \Omega'$ for an element G of $\mathcal{G}_\Omega$, and

$$l_\kappa^\varphi := \sup\left\{\inf_{\Omega'} l_G \mid \varphi + G,\ \varphi + \kappa G \in \mathrm{PSH}(\Omega)\right\}$$

for an element φ of $\Pi_\kappa(\Omega)$.

LEMMA 5.11. *Assume that $\varphi \in \Pi_{\kappa_0}(\Omega)$ and $l_\kappa^\varphi \neq -\infty$ for some $\kappa_0 > 1$. Then for arbitrary strongly pseudoconvex open set $\Omega^* \Subset \Omega$ and $\varepsilon > 0$, there exists a positive number δ_0 such that for δ with $0 < \delta < \delta_0$ and an element v of $L^2_{\varphi+\varepsilon|z|^2}(\Omega^*)$ that satisfy*

$$\begin{cases} \operatorname{supp} v \subset \left\{z \;\middle|\; \dfrac{\delta}{2} < |z_n| < \delta\right\}, \\ \dfrac{\partial v}{\partial \bar{z}_j} = 0 \quad (j = 1, \cdots, n-1), \end{cases} \tag{5.36}$$

there exists an element u of $L^2_{\varphi+\varepsilon|z|^2}(\Omega^*)$ such that

$$\begin{cases} \bar{\partial} u = v\, d\bar{z}_n\,, \\ \|u\|_{\varphi+\varepsilon|z|^2} \leqq C_1 \|v\|_{\varphi+\varepsilon|z|^2}\,, \\ \dfrac{u}{z_n} \in L^2_{\mathrm{loc}}(\Omega^*)\,, \end{cases} \tag{5.37}$$

where C_1 is a constant that depends only on κ_0 and l_κ^φ.

PROOF. Let $\varphi + G$, $\varphi + \kappa_0 G \in \mathrm{PSH}(\Omega)$, and $G \in \mathcal{G}_\Omega$. φ may be taken to be locally integrable. If we define μ_δ by putting $n = 1$ and $\varepsilon = \delta$ in (1.9), then since

$$\partial\bar{\partial}(\log|z_n|^2)_\delta = \frac{1}{2}\mu_\delta(z_n) dz_n \wedge d\bar{z}_n\,,$$

from the assumption, for any $\varepsilon > 0$ we can take a sufficiently small δ such that for any κ with $1 \leqq \kappa \leqq \kappa_0$,

$$\begin{aligned}(5.38)\qquad &\partial\overline{\partial}(\varphi_\delta + \kappa G_\delta + \varepsilon|z|^2) \\ &\geqq \frac{1}{2}\mu_\delta(z_n)dz_n \wedge d\overline{z}_n \quad (z \in \Omega^*).\end{aligned}$$

Set $\eta_A := -G_\delta + A$ $(A > 0)$, and define a quadratic form $Q_{\rho,\Phi}(w)$ to be the one for

$$\begin{aligned}\rho &:= \sqrt{\eta_A + \log \eta_A}\,, \\ \Phi &:= \varphi_\tau + G_\tau + \varepsilon|z|^2 \quad (\tau < \delta).\end{aligned}$$

For such ρ and Φ, we have

$$\begin{aligned}(5.39)\qquad &\rho^2\partial\overline{\partial}\Phi - \partial\overline{\partial}\rho^2 - 4\rho^{-4}\partial\rho \wedge \overline{\partial}\rho \\ &\geqq A(\partial\overline{\partial}\varphi_\tau + \partial\overline{\partial}G_\tau + \varepsilon|z|^2) - \partial\overline{\partial}(\eta_A + \log\eta_A) \\ &\quad - (\eta_A + \log\eta_A)^{-3}\partial(\eta_A + \log\eta_A) \wedge \overline{\partial}(\eta_A + \log\eta_A) \\ &= A(\partial\overline{\partial}\varphi_\tau + \partial\overline{\partial}G_\tau + \varepsilon|z|^2) + \left(1 + \frac{1}{\eta_A}\right)\partial\overline{\partial}G_\delta \\ &\quad + \frac{\partial\eta_A \wedge \overline{\partial}\eta_A}{\eta_A^2} - \left(1 + \frac{1}{\eta_A}\right)^2 \frac{\partial\eta_A \wedge \overline{\partial}\eta_A}{(\eta_A + \log\eta_A)^3}\,.\end{aligned}$$

On the one hand, from the condition $\varphi + \kappa G \in \mathrm{PSH}(\Omega)$ $(1 \leqq \kappa \leqq \kappa_0)$, it follows that $\varphi + \kappa(G - \log|z_n|^2) \in \mathrm{PSH}(\Omega)$. In fact, since

$$\varphi + \kappa(G - \log|z_n|^2) + \alpha\log|z_n| \in \mathrm{PSH}(\Omega)$$

for any $\alpha > 0$, it is enough to let $\alpha \searrow 0$.

Therefore, if δ is sufficiently small, for any τ with $0 < \tau \leqq \delta$,

$$\varphi_\tau + \kappa(G - \log|z_n|^2)_\delta + \varepsilon|z|^2 \in \mathrm{PSH}(\Omega^*).$$

Hence, if we take A so large that

$$\frac{1}{A}\left(1 + \frac{1}{A}\right) < \kappa_0\,,$$

then

$$\begin{aligned}&A(\partial\overline{\partial}\varphi_\tau + \partial\overline{\partial}G_\tau + \varepsilon|z|^2) + \left(1 + \frac{1}{\eta_A}\right)\partial\overline{\partial}G_\delta \\ &\geqq \frac{1}{2}\mu_\delta(z_n)dz_n \wedge d\overline{z}_n\,.\end{aligned}$$

Also, if $A > 4$, it is obvious that the difference between the 3rd and 4th terms in the right hand side of (5.39) is $\geqq 0$.

Therefore, for $w \in C_0^{0,1}(\Omega^*)$,

$$Q_{\rho,\Phi}(w) \geqq \frac{1}{2}\int_{\Omega^*} e^{-\varphi_\tau - G_\tau - \varepsilon|z|^2}\mu_\delta(z_n)|w_n|^2\,dV \tag{5.40}$$
$$\left(w := \sum_j w_j d\overline{z}_j\right).$$

From this, in order to assert the existence of u that satisfies (5.37) for v that satisfies (5.36), it is sufficient to solve the $\overline{\partial}$ equation in $L^{0,q}_{\varphi_\tau+G_\tau+\varepsilon|z|^2}(\Omega^*)$ and let $\tau \to 0$. (Note that C_1 does depend on l_κ^φ as well, because of the appearance of G_τ in the right hand side of (5.40).) □

We use Lemma 5.11 and execute limit operations similar to those in the proof of Theorem 5.10 in order to deduce the following extension theorem:

THEOREM 5.12. *For a pseudoconvex open set Ω and $\varphi \in \Pi_{\kappa_0}(\Omega)$ ($\kappa_0 > 1$), there exists an interpolation operator $I_\varphi : A^2_\varphi(\Omega') \to A^2_\varphi(\Omega)$ if $l_\kappa^\varphi \neq -\infty$. The norm of I_φ depends only on κ_0 and l_κ^φ.*

If $\sup_\Omega |z_n| < \infty$, then $\mathrm{PSH}(\Omega) \subset \Pi_\kappa(\Omega)$ for any $\kappa > 0$, and $l_\kappa^\varphi \geqq -\kappa \sup_\Omega \log|z_n|^2$. Therefore, Theorem 5.12 is a generalization of Theorem 5.10.

A similar method can prove the following extension theorem. (For the proof, see [**36**].)

THEOREM 5.13. *Let Ω be a pseudoconvex open set, and let both φ and ψ be plurisubharmonic functions on Ω. If there exists an element G of $\mathcal{G}_\Omega$ such that $\varphi + G$ is plurisubharmonic and bounded on Ω, then there exists an interpolation operator from $A^2_{\varphi+\psi}(\Omega')$ to $A^2_\psi(\Omega)$.*

This allows us to evaluate the Bergman kernel (Chapter 6) in terms of a geometric invariance of $\partial\Omega$.

5.4. Applications of the L^2 Extension Theorem

5.4.1. Locally Pluripolar Sets. A set of boundary points of a pseudoconvex open set may happen to be removable for functions with low growth rate, such as L^2 holomorphic functions. The problem of characterizing such a function-theoretically small set has been deeply studied in the case of one variable, and in particular, the characterization of removable singularities of bounded holomorphic functions in terms of analytic capacity and that of L^2 holomorphic functions in terms of logarithmic capacity are well-known.

In what follows the latter will be generalized to several variables.

DEFINITION 5.14. We say that a subset E of the complex plane is *locally polar*, or the *logarithmic capacity of* E *is* 0, if for each point x of E, there exist a connected neighborhood U of x and a subharmonic function $\varphi \not\equiv -\infty$ on U such that $E \cap U \subset \{y \in U \mid \varphi(y) = -\infty\}$.

THEOREM 5.15. *For an open set* Ω *in the complex plane and a closed subset* E *of* Ω*, we have* $A^2(\Omega) = A^2(\Omega \backslash E)$ *if and only if* E *is locally polar.*

For the proof, see [**7**] and [**42**].

The concept of local polarity naturally extends to several variables.

DEFINITION 5.16. A subset E of $\mathbb{C}^n$ is said to be *locally pluripolar* if each point x of E has a connected neighborhood U and $\psi \in \mathrm{PSH}(U)\backslash\{-\infty\}$ such that $E \cap U \subset \{x \in U \mid \psi(x) = -\infty\}$.

An analytic subset X of a domain Ω is locally pluripolar. In fact, for a system $\{f_\alpha\}$ of local defining functions of X, it will do to set $\psi := \log \sum\limits_\alpha |f_\alpha|^2$.

The next statement can easily be proved by applying Theorem 5.15 to a function with parameter (the details are omitted).

THEOREM 5.17 (J. Siciak). *If a closed subset* E *of* Ω *is locally pluripolar, then* $A^2(\Omega\backslash E) = A^2(\Omega)$.

As an application of this theorem, we show that *a bijective holomorphic mapping* $F : \Omega_1 \to \Omega_2$ *between domains is biholomorphic.* In fact, if f denotes the Jacobian of F, then F^{-1} is holomorphic on $\Omega_2\backslash F(V(f))$. However, setting

$$\psi(z) := \begin{cases} \log|f(F^{-1}(z))| & \text{for } z \notin F(V(f)), \\ -\infty & \text{for } z \in F(V(f)), \end{cases}$$

since Sard's theorem implies $\psi \not\equiv -\infty$, $F(V(f))$ turns out to be locally pluripolar, and from Theorem 5.17, it follows that F^{-1} is holomorphic on Ω_2.

A generalization of Theorem 5.15 in a rigorous sense is as follows:

THEOREM 5.18. *For two bounded pseudoconvex open sets* $\Omega_1 \supset \Omega_2 \neq \emptyset$*, we have* $A^2(\Omega_1) = A^2(\Omega_2)$ *if and only if for any point* z_0 *of*

Ω_2 and any complex line l through z_0, $l \cap (\Omega_1 \backslash \Omega_2)$ is locally polar in l.

PROOF. Sufficiency is a direct consequence of Theorems 5.15 and 5.10, and necessity is obvious from the same theorems with $\varphi = 0$. □

REMARK. Due to Josefson's theorem, given a pluripolar set E, there exists an element ψ of $\mathrm{PSH}(\mathbb{C}^n)\backslash\{-\infty\}$ such that $E \subset \{z \in \mathbb{C}^n \mid \psi(z) = -\infty\}$. From this, in particular, it follows that a countable union of locally pluripolar sets is locally pluripolar.

5.4.2. Proof of Demailly's Theorem. Siu's theorem has shown a similarity between locally polar sets and analytic subsets. Demailly's theorem, used for the proof of that theorem, is a beautiful application of the L^2 extension theorem as stated below.

PROOF OF DEMAILLY'S THEOREM. Let the notation be that of § 3. 2. First, regarding the sum of the series $\sum_l |\sigma_l(z)|^2$, since $\{\sigma_l\}_{l=1}^{\infty}$ is the orthonormal basis, $\sum |\sigma_l(z)|^2$ coincides with the square of the norm of the following linear function on $A^2_{2m\psi}(\Omega)$:

$$\begin{array}{cccc} \alpha_m: & A^2_{2m\psi}(\Omega) & \longrightarrow & \mathbb{C} \\ & \Uparrow & & \Uparrow \\ & f & \longmapsto & f(z). \end{array}$$

From this by Cauchy's estimate it follows that $\sum |\sigma_l|^2$ converges uniformly on compact sets in Ω and is of class C^ω on Ω, and that the following equation holds:

$$\psi_m(z) = \sup\left\{ \frac{1}{m} \log |f(z)| \;\middle|\; f \in A^2_{2m\psi}(\Omega),\ \|f\|_{2m\psi} = 1 \right\}. \tag{5.41}$$

For $0 < \varepsilon < \delta_\Omega(z)$ and $f \in A^2_{2m\psi}(\Omega)$, since $|f|^2 \in \mathrm{PSH}(\Omega)$,

$$\begin{aligned} |f(z)|^2 &\leqq \frac{n!}{\pi^n \varepsilon^{2n}} \int_{|\zeta - z| < \varepsilon} |f(\zeta)|^2 \, dV \\ &\leqq \frac{n!}{\pi^n \varepsilon^{2n}} \exp\left(2m \sup_{|\zeta - z| < \varepsilon} \psi(\zeta)\right) \int_\Omega |f|^2 e^{-2m\psi} \, dV. \end{aligned}$$

Hence, by taking the supremum of the left hand side within the range $\|f\|_{2m\psi} = 1$, we obtain

$$\psi_m(z) \leqq \sup_{|\zeta - z| < \varepsilon} \psi(\zeta) + \frac{1}{2m} \log \frac{n!}{\pi^n \varepsilon^{2n}}. \tag{5.42}$$

Therefore, the second inequality of (a) is shown.

Next, from n applications of Theorem 5.10 as the dimension increases, it follows that for an arbitrary constant $a \in \mathbb{C}$, there exists an element f of $A^2_{2m\psi}(\Omega)$ that satisfies

$$\int_\Omega |f|^2 e^{-2m\psi}\, dV \leqq C|a|^2 e^{-2m\psi(z)}\,, \tag{5.43}$$

where C is a constant that depends only on n and the diameter of Ω.

If we choose a such that the right hand side of (5.43) $= 1$, then since from (5.41) we must have

$$\psi_m(z) \geqq \frac{1}{m}\log|a|,$$

the first inequality of (a)

$$\psi_m(z) \geqq \psi(z) - \frac{\log C}{2m}$$

is obtained. From this and the definition of the Lelong number, we get $\nu(\psi_m, z) \leqq \nu(\psi, z)$.

It still remains to prove the first inequality of (b). In order to deduce this, since from (5.42), letting C' be the appropriate constant, it follows, in particular, that

$$\sup_{|x-z|<\varepsilon} \psi_m(z) \leqq \sup_{|\zeta-x|<2\varepsilon} \psi(\zeta) + \frac{1}{m}\log\frac{C'}{\varepsilon^n}\,,$$

if we divide both sides of this inequality by $\log\varepsilon$ and let $\varepsilon \to 0$, then the definition of the Lelong number at x implies

$$\nu(\psi_m, x) \geqq \nu(\psi, x) - \frac{n}{m}\,.$$

$\square$

CHAPTER 6

Bergman Kernels

We have already stated several fundamental propositions about the function space $A^2_\varphi(\Omega)$. In this chapter, we will explain the Bergman kernel, which is the reproducing kernel of the space $A^2(\Omega)$. First, we give the definitions and basic facts, and then we prove the boundary holomorphy theorem on biholomorphic mappings between strongly pseudoconvex domains with boundaries of class C^∞. This theorem was obtained by Fefferman, but the proof introduced here is the one due to Bell and Ligocka's idea [**2**], a skillful use of the transformation law of Bergman kernels. Next, a few results on the boundary behavior of Bergman kernels are explained. This is the central problem in the theory of reproducing kernels, but many results on this are beyond the scope of the present book. Hence, we can say that most of the results included here are restricted to elementary cases, and, even so, parts of some proofs are omitted.

6.1. Definitions and Examples

Again, let Ω be a general open set in $\mathbb{C}^n$. For an orthonormal basis $\{\sigma_\mu\}_{\mu=1}^{\infty}$ of $A^2(\Omega)$, Cauchy's estimate implies that the series

$$\sum_{\mu=1}^{\infty} \sigma_\mu(z)\overline{\sigma_\mu(w)}$$

converges uniformly on compact subsets of $\Omega \times \Omega$, and is holomorphic and antiholomorphic on z and w, respectively. This is called a *Bergman kernel function* or *Bergman kernel* of Ω, and is denoted by $K_\Omega(z, w)$.

$A^2(\Omega)$ is a function space whose *reproducing kernel* is K_Ω. That is, for any $f \in A^2(\Omega)$, the value of f at a point $z \in \Omega$ is expressed by the inner product of K_Ω with one variable z fixed and $\overline{f}$, precisely as

$$\begin{aligned} f(z) &= \int_\Omega K_\Omega(z,w) f(w)\, dV \\ &= \left(K_\Omega(z,\cdot), \overline{f(\cdot)}\right). \end{aligned}$$

On the analogy of the significance of the Cauchy kernel $C(z,\zeta) = \dfrac{1}{2\pi\sqrt{-1}}(\zeta - z)^{-1}$ in the case of one variable, we know that in general, analysis of reproducing kernels will bring many good results.

In particular, the Bergman kernel has properties useful in the study of holomorphic mappings, as we will see later, and it is an important research object in the theory of functions of several variables. Since K_Ω is the sum of an infinite series, it is also an interesting object in numerical analysis.

We will give examples of domains in which the exact formulae for K_Ω are given.

EXAMPLE 6.1. In the case $\Omega = \mathbb{B}^n$, as an orthonormal basis of $A^2(\mathbb{B}^n)$, we can take

$$\left\{ \sqrt{\frac{(n + \langle\alpha\rangle)!}{\alpha!\,\pi^n}}\, z^\alpha \right\}_{\alpha \in \mathbb{Z}_+^n}.$$

(The L^2 norms of the z^α may be obtained by induction on the dimension.)

Hence, the Bergman kernel of $\mathbb{B}^n$ is given by

$$\begin{aligned} K_{\mathbb{B}^n}(z,w) &= \sum_{\alpha\in\mathbb{Z}_+^n} \frac{(n+\langle\alpha\rangle)!}{\alpha!\,\pi^n} z^\alpha \overline{w}^\alpha \\ &= \frac{1}{\pi^n}\sum_{\nu=0}^{\infty} \frac{(n+\nu)!}{\nu!} \sum_{|\alpha|=\nu} \frac{\nu!}{\alpha!} z^\alpha \overline{w}^\alpha \\ &= \frac{1}{\pi^n}\sum_{\nu=0}^{\infty} \frac{(n+\nu)!}{\nu!} \langle z,w\rangle^\nu \\ &= \frac{1}{\pi^n}\frac{d^n}{dx^n}\left(\frac{1}{1-x}\right)\Big|_{x=\langle z,w\rangle} \\ &= \frac{n!}{\pi^n}\left(1 - \langle z,w\rangle\right)^{-n-1}. \end{aligned} \tag{6.1}$$

EXAMPLE 6.2. In the case $\Omega = \Delta^n$, from the above calculation and the general formula (which is obvious from the definition of Bergman kernel)

$$K_{\Omega_1\times\Omega_2}((z,z'),(w,w')) = K_{\Omega_1}(z,w)K_{\Omega_2}(z',w'),$$

it follows that

$$K_{\Delta^n}(z,w) = \prod_{j=1}^{n} K_{\Delta}(z_j,w_j) = \frac{1}{\pi^n}\prod_{j=1}^{n}(1-z_j\overline{w}_j)^{-2}\,. \tag{6.2}$$

Besides these, Bergman kernels of various domains have been calculated (see [**29**]).

6.2. Transformation Law and an Application to Holomorphic Mappings

In a case such as $\Omega = \Omega' \times \mathbb{C}$, we have $K_\Omega(z,w) \equiv 0$, and this is not of interest. Hence, we assume below for simplicity that Ω is a bounded domain.

If there exists a biholomorphic mapping $F : \Omega \to \Omega^*$ to another domain Ω^*, then from the integration formula on variable transformation, we obtain the transformation formula for the Bergman kernel:

$$K_\Omega(z,w) = \det\left(\frac{\partial F_j(z)}{\partial z_k}\right) K_{\Omega^*}(F(z),F(w)) \,\overline{\det\left(\frac{\partial F_j(w)}{\partial w_k}\right)}\,. \tag{6.3}$$

Since in particular, for a holomorphic automorphism σ of Ω,

$$K_\Omega(z,z) = K_\Omega(\sigma(z),\sigma(z))\left|\det\left(\frac{\partial \sigma_j}{\partial z_k}\right)\right|^2, \tag{6.4}$$

and $K_\Omega(z,z) > 0$, it follows that $\partial\overline{\partial}\log K_\Omega(z,z)$ is a (1, 1)-form that is invariant under the action of $\operatorname{Aut}\Omega$. In other words, as it is clear that $\partial\overline{\partial}\log K_\Omega(z,z) > 0$, the action of $\operatorname{Aut}\Omega$ is isometric with respect to the Hermitian metric:

$$\sum_{j,k=1}^{n} \frac{\partial^2 \log K_\Omega(z,z)}{\partial z_j \partial \overline{z}_k} dz_j \otimes d\overline{z}_k\,.$$

This metric is called the *Bergman metric* of Ω.

THEOREM (Bremermann's Theorem). *If Ω is complete as a metric space with respect to the Bergman metric, then Ω is obviously a connected domain of holomorphy. Hence, a bounded homogeneous domain in $\mathbb{C}^n$ turns out to be pseudoconvex.*

We will describe an application of the Bergman kernel to holomorphic mappings.

Carathéodory's theorem in the theory of conformal mappings states that if there exists a biholomorphic mapping F between domains Ω_1 and Ω_2 of the complex plane, and if each of $\partial\Omega_1$ and $\partial\Omega_2$ consists of a finite number of simple closed curves, then F can be extended to a homeomorphism from $\overline{\Omega_1}$ to $\overline{\Omega_2}$.[1]

A generalization of this to several variables is the following theorem:

THEOREM 6.3 (Fefferman's theorem). *Assume that there exists a biholomorphic mapping F between strongly pseudoconvex domains Ω_1 and Ω_2 in $\mathbb{C}^n$, and $\partial\Omega_i \in C^\infty$. Then F can be extended to a diffeomorphism of class C^∞ from $\overline{\Omega_1}$ to $\overline{\Omega_2}$.*

PROOF. Let $P_i : L^2(\Omega_i) \to A^2(\Omega_i)$ be the orthogonal projection. If we set

$$u := \det\left(\frac{\partial F_j}{\partial z_k}\right),$$

then from the transformation formula (6.3) we see that

$$u \cdot ((P_2 g) \circ F) = P_1(u \cdot (g \circ F)), \quad \forall g \in L^2(\Omega_2).$$

By the method of indeterminate coefficients, for an arbitrary $h \in C^\infty(\overline{\Omega_2})$, there exists an element v of $C^{0,1}(\overline{\Omega_2}) \cap \operatorname{Dom} \overline{\partial}^*$ such that any derivative of $h - \overline{\partial}^* v$ is equal to 0 on $\partial\Omega_2$, where we apply the method of indeterminate coefficients to the coefficients of the formal power series of the defining functions of Ω_2. (See the proof of Theorem 4.18.)

Therefore, for any $h \in C^\infty(\overline{\Omega_2})$, there exists an element h_0 of $C_0^\infty(\mathbb{C}^n)$ such that $\operatorname{supp} h_0 \subset \overline{\Omega_2}$ and $P_2(h_0 \mid \Omega_2) = P_2 h$.

Regarding an element g of $C^\infty(\Omega_1)$, we write, for simplicity, $g \in C^\infty(\overline{\Omega_1})$ when g can be extended to $\overline{\Omega_1}$ as a C^∞ function.

Since the transformation formula implies

$$u \cdot (h \circ F) = u \cdot ((P_2 h_0) \circ F) = P_1(u \cdot (h_0 \circ F)),$$

in order to show that $F_j \in C^\infty(\overline{\Omega_1})$ $(1 \leqq j \leqq n)$, it suffices to prove that

$$u \cdot (h_0 \circ F) \in C^\infty(\overline{\Omega_1}). \tag{6.5}$$

[1] See Einar Hille, *Analytic Function Theory,* vol. II, Theorem 17.5.3, Ginn and Co., Boston, Mass., 1962.

In fact, in this case, since Kohn's theorem (see Theorem 3.29) implies $P_1(u \cdot (h_0 \circ F)) \in C^\infty(\overline{\Omega_1})$, we have

$$u \cdot (h \circ F) \in C^\infty(\overline{\Omega_1})\,. \tag{6.6}$$

Hence, by putting $h \equiv 1$, we obtain $u \in C^\infty(\overline{\Omega_1})$.

The same argument is applicable to F^{-1}, and in particular, it turns out that u does not have any zero point on $\overline{\Omega_1}$.

Therefore, from (6.6), it follows that $h \circ F \in C^\infty(\overline{\Omega_1})$.

If we use this for $h = z_j$ $(1 \leqq j \leqq n)$, then $F_j \in C^\infty(\overline{\Omega_1})$, and F can be extended to $\overline{\Omega_1}$ as a C^∞ function.

As a similar argument applies to F^{-1}, it follows that F can be extended to a diffeomorphism of class C^∞ from $\overline{\Omega_1}$ to $\overline{\Omega_2}$. □

PROOF THAT $u \cdot (h_0 \circ F) \in C^\infty(\overline{\Omega_2})$. Since F_j is a bounded holomorphic function, by Cauchy's estimate we have

$$|F_j^{(\alpha)}(z)| \leqq c_\alpha \delta_{\Omega_1}(z)^{-\langle\alpha\rangle}, \quad z \in \Omega_1\,,$$

where c_α is a constant that does not depend on z.

Since from $h_0 \in C_0^\infty(\mathbb{C}^n)$ we see that $\operatorname{supp} h_0 \subset \overline{\Omega_2}$, in order to say that every derivative of $u \cdot (h_0 \circ F)$ is bounded, it is sufficient to show that there exists a constant C such that

$$\delta_{\Omega_2}(F(z)) \leqq C\delta_{\Omega_1}(z), \quad z \in \Omega_1\,. \tag{6.7}$$

However, since for a strictly plurisubharmonic defining function r of Ω_1, $r \circ F^{-1}$ is both negative-valued and subharmonic on Ω_2, from Hopf's lemma we get

$$r \circ F^{-1}(w) \leqq -C_1 \delta_{\Omega_2}(w), \quad w \in \Omega_2\,, \tag{6.8}$$

for some constant $C_1 > 0$. (For Hopf's lemma and its proof, see Proposition 6.4 below.)

From (6.8) and the self-evident inequality

$$-r(z) \leqq C_2 \delta_{\Omega_1}(z) \quad \text{for some constant } C_2\,,$$

if we set $C = C_1 C_2$, then (6.7) holds. □

Hopf's lemma used in the above proof is the following proposition:

PROPOSITION 6.4 (E. Hopf). *Let Ω be a bounded domain in $\mathbb{R}^n$ whose boundary is of class C^2, let a be a boundary point of Ω, and let $\nu(a)$ be the inward unit normal line to $\partial\Omega$ at a. Then for an arbitrary negative-valued subharmonic function $u(x)$ on Ω, there exists* *a positive number c such that*

$$\overline{\lim}\,\frac{u(x)}{|x-a|} \leqq -c, \tag{6.9}$$

where the superior limit on the left hand side is taken when $x \in \nu(a)$ and $x \to a$.

PROOF. If we take a point x_0 on $\nu(a)$ sufficiently close to a, there exists an open ball $\mathbb{B}(x_0, R)$ with center x_0 of radius R within Ω whose boundary is tangent to $\partial\Omega$ at a. For a fixed R' with $0 < R' < R$, we can take a sufficiently large positive number λ such that a function

$$v(x) := e^{-\lambda|x-x_0|^2} - e^{-\lambda R^2}$$

is *subharmonic*[2] on $\mathbb{B}(x_0, R) \setminus \overline{\mathbb{B}(x_0, R')}$, and satisfies $v \mid \partial\mathbb{B}(x_0, R) = 0$. If we choose a positive number ε such that

$$\sup\{u + \varepsilon v \mid |x - x_0| = R'\} < 0\,,$$

the maximum principle for subharmonic functions implies $u < -\varepsilon v$ on $\mathbb{B}(x_0, R) \setminus \overline{\mathbb{B}(x_0, R')}$.

From this, (6.9) is evident. □

6.3. Boundary Behavior of Bergman Kernels

The proof of Fefferman's theorem contained in the original article was based on a rigorous analysis of the boundary behavior of Bergman kernels. This would be the main road in the sense of tackling the singular points of reproducing kernels directly, but this much analysis of Bergman kernels requires some treatment of the so-called degenerate elliptic boundary value problems that involve slightly more precise tools than merely the L^2 estimates. In fact, Kohn's theorem used in the above proof is one of those precise tools, and we have no space for them in the present book. However, as the method of L^2 estimates is able to deduce interesting general properties on the singularity of Bergman kernels, we will describe these below.

The next characterization of the value of $K_\Omega(z, z)$ is often used:

$$K_\Omega(z, z) = \sup\{|f(z)|^2 \mid f \in A^2(\Omega),\ \|f\| = 1\}. \tag{6.10}$$

[2]This means $\Delta v \geqq 0$ only for this place.

In fact, $K_\Omega(z,z)$ is nothing but the square of the norm of the following linear mapping:

$$\begin{array}{ccc} A^2(\Omega) & \longrightarrow & \mathbb{C} \\ \Psi & & \Psi \\ f & \longmapsto & f(z) \end{array}$$

Also, since K_Ω is a reproducing kernel, it turns out that the function that realizes the right hand side of (6.10) is the following one:

$$e^{i\theta}\frac{K_\Omega(\,\cdot\,,z)}{\sqrt{K_\Omega(z,z)}} \quad (\theta \in \mathbb{R})\,.$$

It is clear from (6.10) that for an open subset Ω^* of Ω,

(6.11) $$K_{\Omega^*}(z,z) \geqq K_\Omega(z,z) \quad (z \in \Omega^*)\,.$$

Concerning the boundary behavior of Bergman kernels, the following fact is the most fundamental:

THEOREM 6.5. *If $\eta \in C^2(\mathbb{C}^n)$ satisfies $\lim_{z\to 0}|z|^{-2}\,\eta(z) = 0$, and if an open set*

$$\Omega_\eta = \left\{ z \in \mathbb{C}^n \;\middle|\; \operatorname{Im} z_n + \sum_{j=1}^{n-1}|z_j|^2 + \eta(z) < 0 \right\}$$

is pseudoconvex, then

(6.12) $$\lim_{\substack{z\to 0\\ \operatorname{Re} z_n = 0}} K_{\Omega_\eta}(z,z)(\operatorname{Im} z_n)^{n+1} = \frac{n!}{4\pi^n}\,.$$

OUTLINE OF THE PROOF. Since from the Cayley transformation, Ω_0 is biholomorphic to $\mathbb{B}^n$, the transformation formula (6.3) implies

$$K_{\Omega_0}(z,z) = \frac{n!}{4\pi^n}\left(-\operatorname{Im} z_n - \sum_{j=1}^{n-1}|z_j|^2\right)^{-n-1}\,.$$

Hence, (6.12) is equivalent to

$$\lim_{\substack{z\to 0\\ \operatorname{Re} z_n = 0}} \frac{K_{\Omega_\eta}(z,z)}{K_{\Omega_0}(z,z)} = 1\,.$$

Cauchy's estimate implies

$$\lim \frac{K_{\Omega_\eta}(z,z)}{K_{\Omega_0}(z,z)} \leqq 1\,.$$

Now we deduce the reverse inequality. For ζ with Re $\zeta_n = 0$ and $\zeta \in \Omega_\eta$, if we set

$$f_\zeta(z) := \frac{K_{\Omega_0}(z,\zeta)}{\sqrt{K_{\Omega_0}(\zeta,\zeta)}} = \frac{\sqrt{\dfrac{n!}{\pi^n}}\left(\dfrac{\overline{\zeta}_n - z_n}{2\sqrt{-1}} - \langle z', \zeta'\rangle\right)^{-n-1}}{2\sqrt{(-\mathrm{Im}\,\zeta_n - |\zeta'|^2)^{-n-1}}}$$
$$(z' := (z_1, \cdots, z_{n-1}))\,,$$

since $\partial\Omega_\eta$ and $\partial\Omega_0$ contact properly of degree 2 or more at 0, as $\zeta \to 0$, there exists a constant neighborhood U of 0 such that f_ζ is holomorphic on $\Omega_\eta \cap U$, $\|f_\zeta\|_{\Omega_0} = 1$, and

$$\|f_\zeta\|_{\Omega_\eta \cap U \setminus \Omega_0} \to 0\,.$$

Therefore, by solving, with L^2 estimate, the $\overline{\partial}$ equation

$$\begin{cases} \overline{\partial} u = \overline{\partial}(\chi f_\zeta)\,, \\ u(\zeta) = 0 \end{cases}$$

for a C^∞ function χ whose support is contained in U and whose value in a neighborhood of 0 is 1, there exists an element $\widetilde{f}_\zeta$ of $A^2(\Omega_\eta)$ such that $\widetilde{f}_\zeta(\zeta) = f_\zeta(\zeta)$ and $\|\widetilde{f}_\zeta\|_{\Omega_\eta} \to 1$ as $\zeta \to 0$, from which it must follow that

$$\lim_{\substack{z\to 0\\ \mathrm{Re}\, z_n = 0}} \frac{K_{\Omega_\eta}(z,z)}{K_{\Omega_0}(z,z)} \geqq 1\,.$$

□

COROLLARY 6.6 (L. Hörmander, K. Diederich). *Let Ω be a pseudoconvex domain in $\mathbb{C}^n$, and z_0 a strongly pseudoconvex boundary point of Ω. Let r be a defining function of Ω around z_0, and $k(z_0)$ the Jacobian of the Levi form of r at z_0. If $|\mathrm{grad}\, r(z_0)| = 1$, then*

$$\lim_{z\to z_0} K_\Omega(z,z)\delta_\Omega(z)^{n+1} = \frac{n!}{4\pi^n} k(z_0)\,. \tag{6.13}$$

For general domains, no clear relation, as seen above, between the Levi forms of defining functions and the boundary behavior of K_Ω is known. But the next statement is fundamental in a different sense from the above.

THEOREM 6.7. *If Ω is a bounded pseudoconvex domain with boundary of class C^2, then*

$$\underline{\lim}_{z\to\partial\Omega} K_\Omega(z,z)\delta_\Omega(z)^2 > 0\,. \tag{6.14}$$

PROOF. In the case $n = 1$, from the condition we conclude that for each point z_0 of $\partial\Omega$ there exists a circle of constant radius that is contained in $\mathbb{C} \setminus \Omega$ and is tangent to $\partial\Omega$ at z_0. Since the Bergman kernel for the outside of the closed disk satisfies (6.14), from (6.11), (6.14) holds even for Ω. In the case $n \geqq 2$, it suffices to make use of the L^2 extension theorem, Theorem 5.10 (we leave the details to the reader). □

Concerning the Bergman metric, the following is basic:

THEOREM 6.8 (K. Diederich). *If Ω is a pseudoconvex domain, and z_0 is a strongly pseudoconvex boundary point of Ω, then there exist a neighborhood U of z_0 and a positive number C such that for any $z \in \Omega \cap U$,*

$$\partial\bar{\partial} \log K_\Omega(z,z) \lesseqgtr \frac{\partial\delta \wedge \bar{\partial}\delta}{\delta^2} - \frac{\partial\bar{\partial}\delta}{\delta} \pm C\partial\bar{\partial}|z|^2$$

(the double symbols read in the same order), where $\delta := \delta_\Omega$.

For the proof, see [**12**].

A natural question arises on the estimate of the distance function $d(z,w)$ with respect to the Bergman metric:

CONJECTURE. *If Ω is a bounded pseudoconvex domain with boundary of class C^1, then for any $z_0 \in \Omega$, there exists a positive number C such that*

$$d(z_0,z) > \frac{1}{C}|\log \delta_\Omega(z)| - C.$$

At present the following is known in this direction:

THEOREM 6.9. *If Ω is a bounded pseudoconvex domain with boundary of class C^2, then for any $z_0 \in \Omega$, there exists a positive number C such that*

$$d(z_0,z) > \frac{1}{C}\log(|\log \delta_\Omega(z)|) - C\,.$$

For the proof, see [**15**].

For strongly pseudoconvex domains with boundary of class C^∞, the following decisive result is obtained, and there are several studies modeled on this:

THEOREM 6.10 (C. Fefferman [**17**]). *If Ω is strongly pseudoconvex, and if $\partial\Omega \in C^\infty$, then there exist functions φ, $\psi \in C^\infty(\overline{\Omega})$ and a number $\varepsilon > 0$ such that*

$$K_\Omega(z,z) = \varphi(z)\delta_\Omega^{-n-1}(z) + \psi(z)\log\delta_\Omega(z), \quad z \in \Omega\backslash\Omega_\varepsilon\,.$$

Bibliography

[1] U. Angehrn, Y. T. Siu, *Effective freeness and point separation for adjoint bundles.* Invent. Math. **122** (1995), 291–308.

[2] S. Bell, E. Ligocka, *A simplification and extension of Fefferman's theorem on biholomorphic mappings.* Invent. Math. **57** (1980), 283–289.

[3] B. Berndtsson, *The extension theorem of Ohsawa-Takegoshi and the theorem of Donnelly-Fefferman.* Ann. Inst. Fourier (Grenoble) **46** (1996), 1083–1094.

[4] S. Bochner, W. T. Martin, *Several Complex Variables.* Princeton Mathematical Series, vol. 10. Princeton University Press, Princeton, N. J., 1948.

[5] H. J. Bremermann, *Über die Äquivalenz der pseudokonvexen Gebiete und der Holomorphiegebiete im Raum von n komplexen Veränderlichen.* Math. Ann. **128** (1954), 63–91.

[6] ———, *On the conjecture of the equivalence of the plurisubharmonic functions and the Hartogs functions.* Math. Ann. **131** (1956), 76–86.

[7] L. Carleson, *Selected problems on exceptional sets.* Van Nostrand Mathematical Studies, No. 13, D. Van Nostrand, Princeton, N.J., 1967.

[8] H. Cartan, *Les fonctions de deux variables complexes et les domaines cerclés de M. Carathéodory.* C. R. Acad. Sci. Paris **190** (1930), 354–356.

[9] M. Christ, *Global C^∞ irregularity of the $\overline{\partial}$-Neumann problem for worm domains.* J. Amer. Math. Soc. **9** (1996), no. 4, 1171–1185.

[10] J.-P. Demailly, *Regularization of closed positive currents and intersection theory.* J. Alg. Geom. **1** (1992), no. 3, 361–409.

[11] ———, *Complex analytic and algebraic geometry.* Preliminary draft, Institut Fourier, 650 p. http://www-fourier.ujf-grenoble.fr/~demailly/books.html

[12] K. Diederich, *Das Randverhalten der Bergmanschen Kernfunktion und Metrik in streng pseudo-konvexen Gebieten.* Math. Ann. **187** (1970), 9–36.

[13] K. Diederich, J. E. Fornaess, *Pseudoconvex domains: An example with nontrivial Nebenhülle.* Math. Ann. **225** (1977), no. 3, 275–292.

[14] ———, *Pseudoconvex domains with real-analytic boundary.* Ann. of Math. (2) **107** (1978), no. 2, 371–384.

[15] K. Diederich, T. Ohsawa, *An estimate for the Bergman distance on pseudoconvex domains.* Ann. of Math. (2) **141** (1995), no. 1, 181–190.

[16] C. Fefferman, *The Bergman kernel and biholomorphic mappings of pseudoconvex domains.* Invent. Math. **26** (1974), 1–65.

[17] ———, *Parabolic invariant theory in complex analysis.* Adv. in Math. **31** (1979), no. 2, 131–262.

[18] G. B. Folland, J. J. Kohn, *The Neumann problem for the Cauchy-Riemann complex.* Annals of Mathematics Studies, No. 75. Princeton University Press, Princeton, N.J.; University of Tokyo Press, Tokyo, 1972.

[19] J. E. Fornaess, *The disc method.* Math. Z. **227** (1998), no. 4, 705–709.

[20] J. E. Fornaess, B. Stensones, *Lectures on counterexamples in several complex variables.* Mathematical Notes, 33. Princeton University Press, Princeton, N.J.; University of Tokyo Press, Tokyo, 1987.

[21] H. Grauert, *On Levi's problem and the imbedding of real-analytic manifolds.* Ann. of Math. (2) **68** (1958), 460–472.

[22] R. Harvey, *Holomorphic chains and their boundaries.* Several complex variables (Proc. Sympos. Pure Math., Vol. XXX, Part 1, Williams Coll., Williamstown, Mass., 1975), pp. 309–382. Amer. Math. Soc., Providence, R.I., 1977.

[23] F. Hartogs, *Zur Theorie der analytischen Funktionen mehrerer unabhängiger Veränderlichen insbesondere über die Darstellung derselben durch Reihen, welche nach Potenzen einer Veränderlichen fortschreiten.* Math. Ann. **62** (1906), 1–88.

[24] W. V. D. Hodge, *The Theory and Application of Harmonic Integrals.* Cambridge University Press, Cambridge, England; Macmillan Company, New York, 1941 (2nd ed., 1952).

[25] E. Horikawa, *Introduction to Complex Algebraic Geometry.* Iwanami Shoten, Tokyo, 1990 (Japanese).

[26] L. Hörmander, *L^2 estimates and existence theorems for the $\overline{\partial}$ operator.* Acta Math. **113** (1965), 89–152.

[27] ———, *An Introduction to Complex Analysis in Several Variables.* Second revised edition. North-Holland Mathematical Library, Vol. 7. North-Holland, Amsterdam; American Elsevier, New York, 1973 (3rd ed., 1990).

[28] S. Igari, *Real analysis—with an introduction to wavelet theory.* Translated from the 1996 Japanese original by the author. Translations of Mathematical Monographs, 177. American Mathematical Society, Providence, RI, 1998.

[29] M. Jarnicki, P. Pflug, *Invariant Distances and Metrics in Complex Analysis.* de Gruyter Expositions in Mathematics, 9. Walter de Gruyter & Co., Berlin, 1993.

[30] B. Josefson, *On the equivalence between locally polar and globally polar sets for plurisubharmonic functions on $\mathbb{C}^n$.* Ark. Mat. **16** (1978), no. 1, 109–115.

[31] J. J. Kohn, L. Nirenberg, *A pseudo-convex domain not admitting a holomorphic support function.* Math. Ann. **201** (1973), 265–268.

[32] S. Nakano, *On complex analytic vector bundles.* J. Math. Soc. Japan **7** (1955), 1–12.

[33] R. Narasimhan, *Several complex variables.* Chicago Lectures in Mathematics. The University of Chicago Press, Chicago, Ill., 1971.

[34] ———, *Introduction to the theory of analytic spaces.* Lecture Notes in Mathematics, No. 25. Springer-Verlag, Berlin, 1966.

[35] F. Norguet, *Sur les domaines d'holomorphie des fonctions uniformes de plusieurs variables complexes.* Bull. Soc. Math. France **82** (1954), 137–159.

[36] T. Ohsawa, *On the extension of L^2 holomorphic functions. III. Negligible weights.* Math. Z. **219** (1995), no. 2, 215–225.

[37] T. Ohsawa, K. Takegoshi, *On the extension of L^2 holomorphic functions.* Math. Z. **195** (1987), no. 2, 197–204.

[38] K. Oka, *Sur les fonctions analytiques de plusieurs variables. VI. Domaines pseudoconvexes.* Tôhoku Math. J. **49** (1942), 15–52.

[39] ———, *Sur les fonctions analytiques de plusieurs variables. IX. Domaines finis sans point critique intérieur.* Japan. J. Math. **23** (1953), 97–155.

[40] M. Sato, T. Kawai, M. Kashiwara, *Microfunctions and pseudo-differential equations.* Hyperfunctions and pseudo-differential equations (Proc. Conf., Katata, 1971; dedicated to the memory of André Martineau), Lecture Notes in Math., Vol. 287, Springer, Berlin, 1973, pp. 265–529.

[41] B. Shiffman, *On the removal of singularities of analytic sets.* Michigan Math. J. **15** (1968), 111–120.

[42] J. Siciak, *On Removable Singularities of L^2 Holomorphic Functions of Several Variables.* Prace Matematyczno-Fizyczne Wyzsza Szkota Inzynierskaw Radomiu, 1982.

[43] Y. T. Siu, *Analyticity of sets associated to Lelong numbers and the extension of closed positive currents.* Invent. Math. **27** (1974), 53–156.

[44] ———, *The Fujita conjecture and the extension theorem of Ohsawa-Takegoshi.* Geometric complex analysis, World Sci. Publishing, River Edge, NJ, 1996, pp. 577–592.

[45] H. Skoda, *Application des techniques L^2 à la théorie des idéaux d'une algèbre de fonctions holomorphes avec poids.* Ann. Sci. École Norm. Sup. (4) **5** (1972), 545–579.

[46] S. Takayama, *Adjoint linear series on weakly 1-complete Kähler manifolds. I. Global projective embedding.* Math. Ann. **311** (1998), no. 3, 501–531.

[47] ———, *Adjoint linear series on weakly 1-complete Kähler manifolds. II. Lefschetz type theorem on quasi-abelian varieties.* Math. Ann. **312** (1998), no. 2, 363–385.

Index

TITLES IN THIS SERIES

183 **Hajime Sato,** Algebraic topology: an intuitive approach, 1999

182 **I. S. Krasil′shchik and A. M. Vinogradov, Editors,** Symmetries and conservation laws for differential equations of mathematical physics, 1999

181 **Ya. G. Berkovich and E. M. Zhmud′,** Characters of finite groups. Part 2, 1999

180 **A. A. Milyutin and N. P. Osmolovskii,** Calculus of variations and optimal control, 1998

179 **V. E. Voskresenskiĭ,** Algebraic groups and their birational invariants, 1998

178 **Mitsuo Morimoto,** Analytic functionals on the sphere, 1998

177 **Satoru Igari,** Real analysis—with an introduction to wavelet theory, 1998

176 **L. M. Lerman and Ya. L. Umanskiy,** Four-dimensional integrable Hamiltonian systems with simple singular points (topological aspects), 1998

175 **S. K. Godunov,** Modern aspects of linear algebra, 1998

174 **Ya-Zhe Chen and Lan-Cheng Wu,** Second order elliptic equations and elliptic systems, 1998

173 **Yu. A. Davydov, M. A. Lifshits, and N. V. Smorodina,** Local properties of distributions of stochastic functionals, 1998

172 **Ya. G. Berkovich and E. M. Zhmud′,** Characters of finite groups. Part 1, 1998

171 **E. M. Landis,** Second order equations of elliptic and parabolic type, 1998

170 **Viktor Prasolov and Yuri Solovyev,** Elliptic functions and elliptic integrals, 1997

169 **S. K. Godunov,** Ordinary differential equations with constant coefficient, 1997

168 **Junjiro Noguchi,** Introduction to complex analysis, 1998

167 **Masaya Yamaguti, Masayoshi Hata, and Jun Kigami,** Mathematics of fractals, 1997

166 **Kenji Ueno,** An introduction to algebraic geometry, 1997

165 **V. V. Ishkhanov, B. B. Lur′e, and D. K. Faddeev,** The embedding problem in Galois theory, 1997

164 **E. I. Gordon,** Nonstandard methods in commutative harmonic analysis, 1997

163 **A. Ya. Dorogovtsev, D. S. Silvestrov, A. V. Skorokhod, and M. I. Yadrenko,** Probability theory: Collection of problems, 1997

162 **M. V. Boldin, G. I. Simonova, and Yu. N. Tyurin,** Sign-based methods in linear statistical models, 1997

For a complete list of titles in this series, visit the AMS Bookstore at **www.ams.org/bookstore/**.